Petroleum Economics

PETROLEUM ECONOMICS

Issues and Strategies of Oil and Natural Gas Production

Rögnvaldur Hannesson

QUORUM BOOKS
Westport, Connecticut • London

Library of Congress Cataloging-in-Publication Data

Hannesson, Rögnvaldur.
   Petroleum economics : issues and strategies of oil and natural gas
production / Rögnvaldur Hannesson.
     p. cm.
   Includes bibliographical references and index.
   ISBN 1–56720–220–9 (alk. paper)
   1. Petroleum industry and trade. I. Title.
HD9560.5.H26   1998
338.2′7282—DC21   98–6837

British Library Cataloguing in Publication Data is available.

Library of Congress Catalog Card Number: 98–6837
ISBN: 1–56720–220–9

First published in 1998

Quorum Books, 88 Post Road West, Westport, CT 06881
An imprint of Greenwood Publishing Group, Inc.

Printed in the United States of America

The paper used in this book complies with the
Permanent Paper Standard issued by the National
Information Standards Organization (Z39.48–1984).

10 9 8 7 6 5 4 3 2 1

Contents

Preface

This book has grown out of a course in petroleum economics that I have been giving for a number of years to business students at the Norwegian School of Economics and Business Administration. I have been struck by the apparent lack of books that could serve as an introduction to this subject for such an audience. Partly for that reason I decided to gather together the material that I have been using over the years and put it into a book.

The emphasis of this book is on economics and the upstream activities of the oil and gas industry, not on oil refining or retailing of petroleum products. It is aimed at those who want a perspective on energy economics and policy, on questions such as whether or not the world is running out of resources, what is the optimum rate of extraction, what should be done with the oil money, and what are the structural problems associated with a vital and growing petroleum industry. These are important and exciting questions, and the interest in them is by no means limited to those who might have in mind a career in the petroleum industry or the civil service dealing with petroleum and energy issues.

The book begins with a chapter on the development of oil prices. It is now a quarter of a century since the first oil crisis in the fall of 1973. For all persons now over the age of forty that event will stand out in memory. In many countries the oil crisis of 1973 was a watershed in economic development, in some for the better, in others for the worse, although trends unrelated to the oil crises may have been behind structural breaks such as the end of the postwar golden era of economic growth and low unemployment. This chapter ends with a short introduction to futures markets, by now an everyday instrument in the oil business but unknown a quarter of a century ago.

The second chapter discusses changes in the use and supply of energy in greater detail. Modern civilization is founded upon the lavish use of energy stored millions of years ago. It started with coal, and 100 years ago people were worried about the sustainability in the use of coal. Coal was in large part replaced by oil and the worries about sustainability became focused on the availability of oil. Recently the worries about sustainability have shifted from concerns about availability of resources to the possible effects on the global climate from continued emissions of greenhouse gases associated with the burning of fossil fuels. There are as yet no easy and cheap substitutes in sight, except perhaps nuclear energy, and the implications of stabilizing, let alone reducing, the emissions of greenhouse gases are stark, particularly for the poorest and least industrially developed countries of the world.

Chapter 3 deals with the markets for natural gas, with particular emphasis on the market in continental Europe. Use of gas as a fuel is undoubtedly on the rise. Even if it is a fossil fuel it emits much less of greenhouse gases and other polluting substances than coal and oil. The bulkiness of gas sets it apart from crude oil and splits the world market into regional markets with limited interaction. The markets for natural gas have been highly regulated and dominated by national or regional monopolies, but recently the gas markets in the United Kingdom and the United States have been deregulated and made competitive. On the European continent steps are being taken in that direction as well, but because the continental market is dominated by a few large suppliers, most of whom are out of reach for European policymakers, the question remains whether deregulation of the continental European market will bring the same benefits as the deregulation in the United Kingdom and the United States.

Chapter 4 deals with some basic economics of petroleum extraction. The focus is on the tradeoff between the present and the future. Present income is preferred to future income but tilting the income profile toward the present comes at a cost; it requires drilling more wells or constructing more offshore platforms, and may even reduce the total amount of oil that can be recovered from a field. The optimum abandonment time of an oil field is also discussed, and how it might be affected by dismantling costs and the option value of continuing production in the hope of a rising price.

In Chapter 5 the concept of petroleum rent is explained, together with theories of oil price formation associated with Harold Hotelling and Morris Adelman. Despite Adelman's bad luck of having stated, on the eve of the first oil crisis, that the oil price would tend to fall for as far into the future as one cared to look, his theory of the oil price has probably better withstood the test of time and the confrontation with the real world. But no discussion of oil price formation can ignore the Organization of Petroleum Exporting Countries (OPEC). OPEC's pricing policies are discussed in the light of the leader-follower model and how the cartel may have misjudged the long-term elasticity of demand and the response of other oil producers to high oil prices. Finally, attention is drawn to the shift in taxation of oil rents away from producing to consuming countries that appears to have taken place in recent years.

Chapter 6 considers oil taxation in greater detail. The petroleum industry has long been a bountiful source of revenue for governments in various oil-producing countries and provinces. Even if taxing pure rents should not affect the efficiency of the industry being taxed or the economy at large, it has been easier said than done to design tax systems that actually are neutral in this sense and only tax away pure rents. The discussion of this issue draws on petroleum taxes in Norway and their development over time.

But why tax oil rents, and how should the tax revenue be spent? Oil, like other minerals, is a nonrenewable resource. Oil rents are incomes from oil extraction, over and above what is needed to get the oil out of the ground. The price of oil is so much higher than the cost of production for most oil finds that these rents represent a substantial source of wealth for oil-producing countries, states, or provinces. Spending this wealth on current consumption can substantially enhance the standard of living for the people of those areas, but once it has been spent in this way it will be gone forever. A more forward-looking way of spending the oil rents and one that would be more equitable across generations is to transform the oil rents into productive capital that can benefit not just the present but also future generations. Some oil-producing countries and provinces have channeled a part of their oil rents into investment funds specially set up for this purpose. This issue is discussed in Chapter 7. Some people take the view that governments cannot in fact be trusted with this task owing to the shortsightedness implicit in the electoral cycle, and that the oil wealth therefore had better be stored in the ground and depleted gradually to satisfy current needs at any point in time. Storing oil wealth in this way is not very productive, however, unless the price of oil is rising over time. But prices can fall as well as rise, even the price of oil, so how does this kind of uncertainty affect the degree to which oil should be extracted immediately or left in the ground? The answer to this question is ambiguous, even if we assume that societies, or the governments acting in their trust, are risk averse.

A sudden discovery of petroleum resources, or a substantial appreciation of the value of such resources, poses formidable adjustment problems for the countries or provinces affected. Some of the newfound riches will be spent on current consumption, even if due account is taken of future generations. But enhanced well-being will affect different industries disproportionately and some might even decline. These adjustment problems are discussed in the final chapter. In the 1980s these problems were often referred to as the Dutch Disease, but the Dutch economy seems currently to be in a healthier shape than the economy of many, if not most, other countries in western Europe. Sweden, a country not plagued by sudden and substantial discoveries of natural resources, seems to have symptoms similar to Norway, which over the last quarter of a century has grown from zero oil exports to the second largest oil exporter in the world. Perhaps the role of natural resource wealth in economic and political stresses of resource-rich countries has been exaggerated.

Some friends and colleagues have been kind enough to read parts of the manuscript and provide comments. Prof. Eirik Amundsen, of the University of Bergen, took the trouble to read the entire manuscript and comment extensively. Dr. Arild Nystad, previously at the Norwegian Petroleum Directorate, Anton Hellesøy of Norsk Hydro, Jan Bygdevoll of the Norwegian Petroleum Directorate, and Dr. Øystein Thøgersen of the Norwegian School of Economics and Business Administration, all read parts of the manuscript and provided comments. I am grateful to them for their endeavors, and they are all exonerated from whatever errors, mistakes, and misplaced emphasis that might remain.

Chapter 1

Oil and Oil Prices

OIL: ITS VITAL ROLE

Oil in modern society is like blood in the human body. Without it modern society as we know it would cease to exist. Blood flows through our arteries carrying energy to our muscles and brain, allowing us to function. Oil flows through pipelines carrying the energy that allows the wheels in our machines and transportation equipment to go round and makes our dwellings more comfortable in too hot or cold weather. Whenever the supplies of oil have been seriously interrupted the inconveniences have been immediate, and people have adopted smart and less smart contingencies to cope with the situation. Many still remember the lines outside gas stations in California and elsewhere in the United States during the Arab oil embargo in the wake of the Yom Kippur War[1] in 1973. Rationing of gas became the order of the day throughout the Western world. I remember meeting someone who got a can of gas as a Christmas present from his parents that year. A Swedish newspaper reported that someone had stored gas in his bathtub.

It was not always so. The use of oil as a source of energy is a recent phenomenon. The beginning of the oil industry is usually dated to 1859 when in Pennsylvania an oil well was struck instead of brine (Jones, 1988), but oil seepages have been known and taken advantage of for much of human history and in as various parts of the world as China, the Roman Empire, and Brazil in precolonial times (Van Meurs, 1981). The word ''naphtha'' is of ancient Greek origin.

Oil has distinct advantages as a carrier of energy. It has a high content of energy per weight unit, which minimizes transportation costs. It is rather easily handled, being fluid and storable without too much cost. Oil quickly became a preferred carrier of energy. Within a few years it replaced whale oil in street lighting. But the internal combustion engine was the decisive factor. The fluidity of oil made uninterrupted supply of fuel a relatively simple thing, and the high energy-to-weight ratio made it possible to carry sufficient energy supplies in a vehicle for travel over long distances.

The oil industry and the uses of oil developed first and foremost in the United States, and it was one of the main engines of the rapid industrialization and growth of the American economy in the late nineteenth and early twentieth century. Some of the great American fortunes, inter alia that of the Rockefeller family, were made from oil. Oil quickly became important enough to shape world political events. Intense rivalry developed between the Germans and the British in the late nineteenth century in their diplomatic approaches toward Turkey, but Turkey then held sway over the Middle East, parts of which appeared to be promising for oil exploitation. After World War I and the disintegration of the Turkish empire Iraq and Kuwait, soon to emerge as important oil producers, came under British influence. The Shah of Iran was put in power by the British and American governments because of interruptions in oil supplies from Iran under the government of Mossadeq in the early 1950s. The Iraqi invasion of Kuwait in 1990 was rolled back, mainly through the efforts of the United States, undoubtedly because of the threat it represented to the supplies of oil from the Middle East.

The success of oil would never have come about unless it was available in sufficient quantities at a reasonable price. That this should be so is by no means self-evident and indeed somewhat counterfactual. Oil is a nonrenewable resource. It is formed underground from remains of organic material such as debris from marine organisms and one-time forests.[2] This is an ongoing process but the rate of formation of new oil is so slow as to be of negligible interest; what we are using up now is immensely greater than whatever new oil is being formed. Over time we will run out of oil, but nobody knows how large the reserves are on which we can draw. Ever since oil was first discovered, the additions to our known reserves have been as great as or greater than what we are using up, keeping the supplies abundant. There is no end yet in sight of this affluence; at the present time (1997) the known reserves are sufficient to cover our current use for more than forty years (*BP Statistical Review of World Energy*, 1997). This does not mean that we will run out of oil in forty years; new oil finds have continuously been discovered to keep reserves ahead of production. For decades worldwide proven reserves of oil have covered more than thirty years at the contemporary rate of extraction.

OIL PRICES SINCE WORLD WAR II

Anyone looking for evidence of an increasing scarcity of oil will be hard put to find it. Additions to known reserves have by no means always come from finds such as offshore fields that are expensive to exploit; discoveries of inexpensive finds are still being made, in the Middle East and elsewhere, and technological progress exerts a steady downward pressure on extraction costs. For these reasons we would expect a falling rather than a rising price of oil. Nevertheless the price of oil has shown a peculiar behavior over time. Figure 1.1 shows how the price of oil has developed since the early beginnings of the industry, in nominal and real terms. Early on the price of oil fluctuated violently and was in fact well above the 1980 peak in real terms. The 1950s and 1960s were a period of relatively stable and declining real prices of oil. The early 1970s, particularly 1973–74, mark a watershed. Since then the oil price has been subject to three major shifts; it rose dramatically in 1973–75 and again in 1979–81, and fell steeply in 1986.[3]

As is to be expected for such a vital commodity as oil, the price gyrations of the 1970s and 1980s caused major disruptions and concerns. The two rises represented enormous changes in terms of trade, to the benefit of oil-exporting countries and the disadvantage of oil-importing countries. The balance-of-payments problems and the necessary adjustment that followed were formidable, especially for those who were on the losing side of these price changes. The postwar golden age of economic growth in the Western world came to an end.[4] Newspapers and periodicals were flooded with articles on these matters, some of which took the view that the price of oil was something unique and unaffected by the laws of economics.[5]

Even if changes in the price of oil sometimes, and for a limited period, may appear contrary to the laws of economics, this is hardly so on closer inspection, but the way the laws of economics make their impact on the oil industry can be highly idiosyncratic. There are a few peculiarities that are important for understanding the movements in the price of oil. First, the oil industry is dominated by a few large companies. This was particularly true in the past when the oil companies cooperated in setting their prices. Second, because oil is a storable commodity, the oil market is highly influenced by expectations about future prices and changes in inventories. This in particular may account for changes in oil prices that appear counterintuitive and counterfactual.

The steady decline in the oil price that persisted for much of the first quarter of a century after the end of World War II took place despite a steadily rising demand for oil. The years up to the early 1970s were a period of sustained and unparalleled economic growth in the industrialized world. This growth was accompanied by an even greater rise in the demand for oil. That the price of oil nevertheless fell was due to two factors. First, as already mentioned, sufficient oil was discovered to replace the increasing quantities that were used up. Second, the new reserves that were found were to an increasing degree controlled by so-called "independent" oil companies, independent, that is, of the international giants that controlled the great reserves in Saudi Arabia and elsewhere and cooperated in exploiting them.[6]

The history of price fixing among the large oil companies goes back to well before World War II. A story is told of a meeting between the top leaders of

Figure 1.1
The Price of Crude Oil 1861–1997

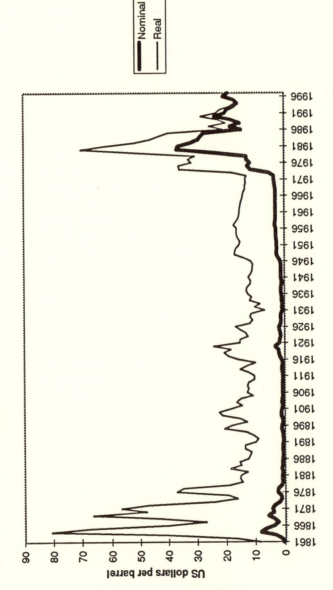

Note: 1861–1972: wellhead prices in the United States; 1973–79: Saudi Arabia posted price or official wellhead price; 1980–97: Brent Blend spot price. Real prices are nominal prices inflated to 1996 dollars by the *US Consumer Price Index*.

Sources: Statistics Norway, *1986 North Sea Oil and Gas Yearbook*, and *Oil and Gas Activity*; U.S. Bureau of the Census, *Statistical Abstracts of the United States*, and *Historical Statistics of the United States, Colonial Times to 1970*.

Esso and BP at the Scottish castle of Achnarry in 1928 where the elements of cooperation on prices and market shares were agreed upon (Jones, 1988; Adelman, 1995). A major component of this agreement was the so-called point pricing system, according to which the price of oil should be fixed as the price in the Mexican Gulf plus transportation costs. As oil increasingly came from elsewhere, the Middle East in particular, the collusive character of this arrangement became increasingly obvious. Over time a two-point pricing system developed, with the other point being in the Persian Gulf. The growth of the so-called independent oil companies, which did not feel bound by this agreement, put the point pricing system under increased strain and maintained a downward pressure on prices.

THE RISE AND FALL OF OPEC

Around 1970 there were signs that the era of slowly falling real oil prices was coming to an end. The price rose in the early 1970s, which some people thought was due to a conspiracy of the oil companies. But an important structural change had been taking place in the world oil trade. Even if more oil was found than was being used up, these discoveries were distributed in a lopsided fashion. Most of these discoveries and, no less importantly, the ones that were easiest to exploit were located in the Middle East. The countries in western Europe imported most of their oil from the Middle East, and the United States was becoming increasingly dependent on imports from this source. Consequently, disruptions in supply from the Middle East could be expected to have a major impact on the world oil market. The Organization of Petroleum Exporting Countries (OPEC)[7] was becoming increasingly aware of this and was adopting a more aggressive stance in its negotiations with the international oil companies about "posted prices," on which the tax payments to the governments of the oil-producing countries were based.

But it was the war between Israel and Egypt in October 1973 that precipitated the dramatic increase in oil prices. The Arab countries imposed an oil embargo on the United States and the Netherlands in the wake of the war, in order to put pressure on Western governments because of their support of Israel. Suddenly it became evident how dependent the Western industrialized countries had become on oil imports from the Arab world. Gasoline was rationed through various means in the oil-importing countries, and in some countries elaborate plans for oil rationing were made.

The oil embargo was over quickly, but what had become evident was the power the oil-exporting countries wielded over the price of oil. The price of oil in the international oil market, which at that time was mainly a fairly informal set of deals concluded in bars and restaurants in Rotterdam, shot up in the wake of the embargo, making it evident what the oil-exporting countries could achieve by withholding supplies. The oil-exporting countries quickly made use of this opportunity. For years OPEC had negotiated so-called posted prices with the oil companies. These prices had little to do with the actual prices of oil in international trade but were merely reference prices used to calculate the tax revenues of the oil-exporting countries. Now these posted prices, together with the fact that the host governments nationalized the international companies extracting oil on their soil, became prices valid in transactions between the producing countries and the international oil companies and others who bought the oil.

At this time and well into the 1980s the world held its breath before every meeting of the oil ministers of the OPEC countries. What price would they come up with this time? Many appeared to believe that OPEC just had to name a price; the market would take anything they would come up with.

How could this happen? The key to the answer is the limited responsiveness of demand for oil in the short term to changes in its price (low price elasticity of demand).[8] This gives suppliers a large leeway to extract the price they see fit when acting in concert. At the time there was much debate as to what governed OPEC's behavior, to what extent it was a cartel, and so forth (see, for example, Griffin and Teece, 1982). Those who contested the hypothesis that OPEC really was a cartel pointed to the fact that it did not set production quotas, something any effective cartel will have to do. Yet the hypothesis of the absence of cartel power was difficult to reconcile with the apparent sway OPEC held over oil prices.

The most reasonable answer to the riddle that OPEC appeared to control oil prices and yet did not set any production quotas appears to be that the organization was for a time a victim of its own success. It could climb, as it were, up a steep demand curve. The response on behalf of buyers to the rising price was limited. There are few good substitutes for oil products, and wherever there are substitutes it takes time to put them to use. There is no alternative to oil products in aviation, and few in ground or ocean transport. The scope is greatest in the generation of electric energy, but in these cases it is often necessary to convert power stations to run on coal or other fuels, or to build new ones. The main vehicle for limiting the demand for oil products in the short run is cutting down on transportation and the production of electric energy, as indeed happened; the two oil price rises in the 1970s triggered worldwide recessions. There was, therefore, limited need for OPEC to set production quotas, as total production did not have to be cut by much to support a higher price. This was most likely reinforced by the fact that the need for the OPEC countries to export oil to obtain foreign currency became less as the price of oil went up; the capacity to absorb imports lagged behind the sudden and substantial increases in foreign currency earnings.

This, however, did not last forever. The growth in oil demand slowed immediately after the first oil price rise. Figure 1.2 shows the total world production of oil since 1945 on a logarithmic scale. The curve is almost a straight line until the first dramatic price rise (1973–74), indicating a constant rate of growth (about 7 percent per year). Since then, and particularly after the second price rise around 1980, the demand for oil has stagnated; in 1994 it was virtually the

Figure 1.2
World Production of Crude Oil since World War II (Logarithmic Scale)

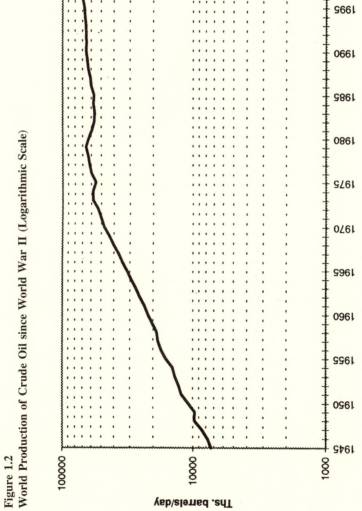

Sources: Statistics Norway, *1986 North Sea Oil and Gas Yearbook, Petroleum Economist, and BP Statistical Review of World Energy.*

same as in 1979. The fact that the demand for oil continued to rise for a while in the 1970s, after an initial drop in the wake of the first price rise, may be due to the time that normally lapses before engines and power plants are changed or replaced for the purpose of using substitutes, or for using oil products more frugally, rather than a further price rise being necessary to bring about such changes.

The years from the late 1970s to late 1985 can be characterized not by OPEC as a successful cartel but rather by Saudi Arabia acting as a swing producer, that is, a producer that tempers its own production as needed to maintain the price. The Saudis had been among the least militant in terms of keeping the price high, apparently concerned with the long-term effect of such a policy.[9] When the Iranian production of oil fell dramatically after the revolution of 1979, Saudi Arabia increased its own production to make up for the shortfall, in order to relieve the pressure on prices. Nonetheless prices rose to an all-time high of over US$ 40 per barrel, a development apparently associated with the buildup of precautionary inventories and expectations of a further rise. But after the initial panic reaction to the events in Iran had worn off, prices began to slip. Supplies from non-OPEC sources, mainly the North Sea and Mexico, increased rapidly, in part because the high oil price had made high-cost areas such as the North Sea very profitable. This put a further downward pressure on prices and made it necessary to cut production in order to maintain the price to which OPEC committed itself. In 1982 production quotas could no longer be avoided. The agreements on quotas were not reached easily, and some OPEC members have been prone to exceed their quotas. Saudi Arabia was, for a time, prepared to cut its own production to maintain the agreed-upon price.

As a result of its role as a swing producer, the stagnant demand for oil, and the increase in supplies from non-OPEC sources, the oil production of Saudi Arabia shrank from about 10 mill. barrels a day in 1979–80 to about 3 in 1985. The impact on the Saudi export revenue was even more dramatic; as oil prices fell from over $40 per barrel at their peak in 1980 to less than 30 in 1985 the Saudi revenue from oil exports fell from 119 billion dollars in 1981 to 27 in 1985. From having one of the world's largest surpluses on current account, Saudi Arabia went to having the world's second largest deficit after the United States, clearly not a sustainable position.

In late 1985 Saudi Arabia changed its strategy and stopped supporting an oil price of $28 per barrel by restraining its own production as necessary to compensate for other OPEC members' unwillingness to cut their own production. Instead of selling only at a predetermined price, the Saudis started to engage in so-called netback deals, by which the oil price is derived from the market price of petroleum products less refining costs. The Saudi Arabian exports started to pick up, and so did the total supply of oil. In early 1986 the market was flooded with oil and the price dropped as precipitously as it had risen in the 1970s and early 1980s, hitting a low point close to $10 per barrel in the summer of 1986, a drop by more than one half from a year earlier. Whether supply exceeds or falls short of normal demand, the low short-term price elasticity of demand causes dramatic price movements, but down this time instead of up.

Since 1986 the price of oil has been variable but has mainly stayed between $15 and $20 per barrel. The supply disruption following the invasion of Kuwait in 1990 had a less dramatic impact on prices than the oil embargo of 1973 and the Iranian revolution in 1979; in the fall of 1990 the oil price briefly exceeded $30 per barrel. The years since 1986 have been characterized by the waning power of OPEC; the organization has not succeeded in maintaining an oil price of $20 per barrel as has been its goal. Two factors contribute to OPEC's weakness: the fall in its market share and the strains on the internal discipline of the organization. In the 1970s OPEC's share of world production was more than 50 percent, but as the North Sea and other new areas came on stream its share fell rapidly, reaching a low point of about 30 percent in 1985. By 1996 it had risen again to 40 percent but this is apparently not enough to give the organization a sufficient market power.

The low price and the spending habits that the OPEC countries developed after the gains they made in the 1970s also put a heavy strain on the internal discipline of the organization. In the 1970s when oil prices were rising and the foreign currency expenditures of the OPEC countries lagged behind their earnings it was easy to accept lower production quotas; more money was not urgently needed, and lower quotas, if adhered to by all, meant more and not less money. Now the situation is quite different. Most OPEC members need more foreign currency, some desperately. Lower production quotas mean higher incomes only if all abide by them, whereas producing a bit more than one's quota will increase the revenue for the country that does so, and gets away with it, almost in proportion to the excess production, while most of the cost is borne by others through a lower price suffered by all. Besides, the experience since the mid-1970s calls into question the long-term wisdom of pushing for higher oil prices; in the end it accomplished little for the OPEC members; it encouraged production elsewhere and halted the rising demand for oil. As this book was being finished (early 1998), the real price of oil was not much higher than before the first oil price rise in 1973–75.

A NEW ROLE OF MARKETS

Even if the dramatic events in the wake of the Yom Kippur War in 1973 have not had a lasting effect on the real price of oil they did change the workings of the oil market in a fundamental way. The nationalization of the oil-producing operations in the producing countries of the Middle East, together with the continued strengthening of the "independent" oil companies, has robbed the "majors" of their previous control of the oil market. Instead of flowing within a vertically integrated company, oil that is produced in Saudi Arabia, Iran, and other countries where the "majors" used to have concessions now must be marketed by the national oil companies of these countries. Oddly, the nationalization of the oil companies in the Middle East may have increased the role of markets by strengthening the need for organized exchanges, but the emergence of new production areas such as the North Sea has also contributed to this.

Nowadays oil is sold like any other commodity on commodity exchanges in various places around the world, such as the International Petroleum Exchange (IPE) in London and the New York Mercantile Exchange (NYMEX). Oil can now be sold "spot," i.e., for immediate delivery, or forward, for delivery at a later date. A special case of forward markets and by far the most important are so-called futures markets where contracts for delivery after a specified number of months are traded. The spot price of oil, and the prices of the futures contracts, are quoted daily on the business pages of the newspapers. Whereas earlier there was considerable doubt as to what really constituted an arm's-length price between independent agents, this information is now readily available from any major newspaper.

On futures markets most of the dealings, typically about 99 percent, are "in paper"; that is, the contracted oil is never delivered. Instead the deal is reversed before the actual delivery is to take place; i.e., the contract is bought back and nullified. The point of such deals is to hedge against price changes, and to make a profit by buying cheap and selling dear. Futures contracts are standardized with respect to what kind of oil is to be delivered and where. This is necessary in order to achieve a large volume in the market, a prerequisite for smooth price changes, and to ensure convergence between the futures price and the spot price. Even if one deals exclusively in "Brent Blend" on the International Petroleum Exchange in London and in "West Texas Intermediate" on the New York Mercantile Exchange this can still be attractive for those who deal in other types of crude oil, because the prices of different types of crude oil typically move in parallel, the difference being determined by quality differences such as sulfur content and whether the oil is light or heavy. The delivery point has to be chosen so as to be attractive for some critical mass of market participants, in order to ensure convergence of spot prices and prices in futures contracts for which the delivery date is approaching.

The usefulness of futures contracts as a hedge against price variations lies in the parallel movement of prices of contracts with different dates of maturity and the convergence of futures prices and spot prices. Table 1.1 shows midmonth quotations of futures prices of crude oil on NYMEX for four futures contracts, one maturing in June, another in July, a third in September, and a fourth in November 1997. Except for the March-to-April change, all prices move roughly in parallel, with a high point in May. A buyer who in March planned to buy oil in May and feared that the price would rise could have secured for himself the low price prevailing in March by buying, say, a September contract for $19.43 per barrel in March. In May he would have sold that contract for 21.02 and made a gain of $21.02 - 19.43 = 1.59$ on the futures contract transactions.

Table 1.1
Prices for Futures Contracts on NYMEX, U.S.\$ per Barrel

Month of quotation	Contract maturing in			
	June 97	July 97	Sept. 97	Nov. 97
March 97	20.91	19.49	19.43	
April 97	19.85	19.88	19.95	
May 97	21.30	21.26	21.02	20.82
June 97		19.07	19.34	19.58
July 97			18.10	18.35
August 97			20.08	20.31

Note: Quotations at midmonth.

Provided the spot price in May was close to the quotation for the soon-maturing June contract, he could have bought physical oil spot at about 21.30 and realized a price close to the March price; if we subtract the gain from the futures contract transactions from the spot price of oil we arrive at a realized price of 21.30 − 1.59 = 19.71. Even after adding broker fees this is probably a better deal than he could have got by buying at the spot price in May.

We could have told the same story for a seller of crude but would then have come up with the sad conclusion that he would have managed to secure for himself a lower price than that prevailing in May. But the point is to hedge against price changes; prices move up and down, and if one prefers certainty over uncertainty one must be prepared to live with missing some opportunities of making a gain from favorable price movements. Who, then, picks up the tab? So-called speculators engage in futures trading for the purpose of making a gain from favorable price movements and assume the price risk that buyers and sellers wish to unload. The basis for this gain is that a buyer who wants to avoid price risk is willing to buy for future delivery at a price that is somewhat higher than he expects will prevail on the day he in fact buys the product and, similarly, a seller is willing to sell for future delivery at a price that is lower than the price he expects to prevail in the future. Whether or not futures prices will be higher or lower than spot prices is difficult to say, however; it depends on who most want to hedge, buyers or sellers, and this is likely to be related to the direction in which prices are generally expected to move. In Table 1.1 we see that the prices of contracts that are close to maturity (and hence close to the spot price) are sometimes higher and sometimes lower than prices of contracts

Figure 1.3
Prices of Oil Futures (Brent Blend) on the International Petroleum Exchange in London for Contracts Maturing in January to May 1998

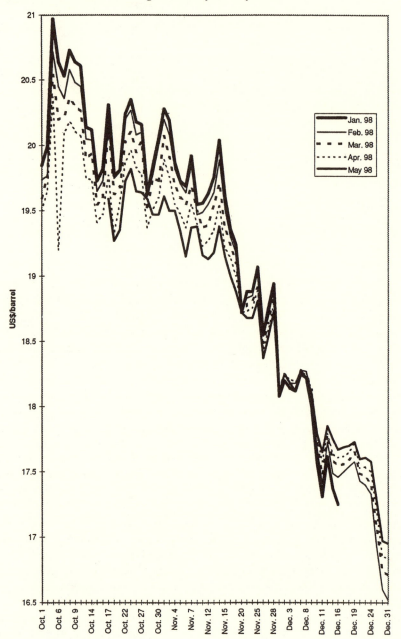

Source: Financial Times.

that mature at a later date. In April 1997 the price of the June contract was lower than that of the July contract, which again was lower than that of the September contract, but in May the order had been exactly reversed. In June the order prevailing in May had again been reversed. But generally the prices of futures contracts move in parallel, with the price of contracts usually being higher the further into the future they mature.

This last phenomenon can be explained by considering the alternative to price hedging in the futures market, which is buying the commodity and storing it. But storing costs money, both directly and indirectly through the opportunity cost of the money spent on buying the commodity (the money could have been invested in an asset that yields some return). It pays rather to use the futures market as long as the price of the contract does not exceed the spot price by more than the cost of storing the commodity, but the longer the commodity is stored, the higher the cost. The opportunity cost of buying and storing the commodity sets an upper limit to how much the price of a futures contract can rise with the time to maturity but the opposite is perfectly possible; the price of futures is sometimes lower the longer the time to maturity. Figure 1.3 shows the day-to-day development of futures prices on the International Petroleum Exchange in the last quarter of 1997. The futures prices move nicely in parallel, but until about mid-November the price was lower the longer the time to maturity. The reason could have been that the price was generally expected to fall, and the trend is indeed downward. In late November and early December the prices of all contracts were almost identical, and in December the more normal situation emerged where the price rises with the time to maturity.

NOTES

1. So called because Egyptian forces crossed the Suez Canal and attacked Israel on the eve of Yom Kippur (the Day of Atonement), one of the main Jewish holidays. Syria joined Egypt in the war, but the two ultimately lost.

2. This, at any rate, is the dominant theory in Western countries. In the former Soviet Union an entirely different theory is dominant. This theory was put forward in the 1950s and rapidly gained ground in that part of the world. According to this *abyssal, abiotic* theory, oil is formed under tremendous pressure deep in the earth's crust from primordial material and has erupted toward the surface where it is found in reservoirs underneath layers of solid rock. If true, this would mean that there is practically no limit to how much oil can be extracted, and it would also explain why oil is found in places where it had been thought unlikely to occur. On this, see J. F. Kenney in *Energy World* (June 1996): 16–18, a rejoinder from Shell in the same, March 1997, p. 14, and, as of the time of this writing, an unpublished reply by Kenney, Gas Resources Corporation, Houston, Texas.

3. The story of oil prices since 1945 is vividly told in Adelman (1995).

4. In most OECD countries the rate of GDP growth was much higher before the first oil crisis in 1973 than after. Typical numbers for the annual average growth rate of GDP per capita are as follows:

	1950–73	1973–95
United States	3.6	2.2
United Kingdom	3.0	1.6
France	5.0	1.9
Sweden	4.0	1.4
Norway	4.1	3.6

Only oil-rich Norway has had a growth rate after 1973 comparable to what went before. On the other hand, it could be argued that all these countries, and the rest of the most advanced OECD countries, have gradually entered a state of maturity in which less rapid growth rates are to be expected irrespective of oil prices. Norway was the least developed of these five in 1973, and the appearance of the petroleum sector would in any case have propelled her economic development. *Source*: Maddison (1991) and calculations based on OECD, *Historical Statistics 1960–95*.

5. Milton Friedman, of the University of Chicago, was one of the dissenting voices, writing in his column in *Newsweek* (February 17, 1975): "Almost regardless of our energy policy, the OPEC cartel will break down. That is assured by a worldwide reduction in crude-oil consumption and expansion in alternative supplies in response to high prices. The only question is how long it will take."

6. These giants used to be known under the nickname of the Seven Sisters. They consisted of Mobil, Standard Oil of California, Gulf, Texaco, Esso, BP, and Shell. In 1984 Standard Oil of California and Gulf merged. One of the consortia formed to exploit joint "concessions" in the Middle East was Aramco, which operated in Saudi Arabia and was jointly owned by Esso, Texaco, Mobil, and Standard Oil of California. The share of the Seven Sisters in the international trade of oil fell from 98 percent in 1950 to 89 percent in 1957, and further to 76 percent in 1969.

7. OPEC was founded in 1960 by Iran, Iraq, Kuwait, Saudi Arabia, and Venezuela, for the purpose of increasing the tax income of the producing countries. Members of OPEC as of 1997 are Saudi Arabia, Iran, Kuwait, United Arab Emirates, Iraq, Qatar, Indonesia, Nigeria, Libya, Algeria, Gabon, and Venezuela. Equador left the organization in 1993 and Gabon did so in 1995.

8. Moran (1982, p. 102) reports values between 0.2 and 0.9.

9. All would not agree on this; see, e.g., Adelman (1995).

Chapter 2

The Markets for Energy

A HISTORICAL PERSPECTIVE

The demand for oil is the result of our demand for energy. Oil is burned in engines that move axles and wheels propelling our transportation equipment or making our production equipment perform useful tasks; it is burned in furnaces to generate heat; and it is used to drive generators producing electricity, which in turn heats our houses, cooks our meals, drives our production equipment, etc. But oil is not our only source of primary energy. Indeed it is a latecomer to the scene. Coal was the primary source of energy during the industrial revolution, and before that the main source of energy was wood or peat, domesticated animals, and the human body itself. In the late nineteenth and early twentieth century some people were as worried that the world would run out of coal as others were in the 1970s and 1980s that it might run out of oil.[1]

During much of the twentieth century the share of oil as a source of energy increased steadily at the expense of coal. The falling real price of oil made it increasingly competitive vis-à-vis coal, and the coal-producing countries of western Europe tried to protect their ailing coal-mining industries directly and indirectly through high taxes on oil products and subsidies to coal production but managed only to slow down the process of decline. Figure 2.1 shows the shares of oil, natural gas, and coal in the consumption of energy since 1950. The share of oil and natural gas increased without interruption until the first oil price hike in 1973–75. From the mid-1970s to the mid-1980s, the era of high oil prices, the share of oil fell substantially, but has remained stable since the dramatic fall of oil prices in 1986. The share of natural gas has continued to rise, and coal has not recaptured its previous position. Figure 2.2 shows the different sources of primary energy and includes hydroelectric and nuclear

Figure 2.1
Shares of Different Fossil Fuels in World Consumption, 1950–1997

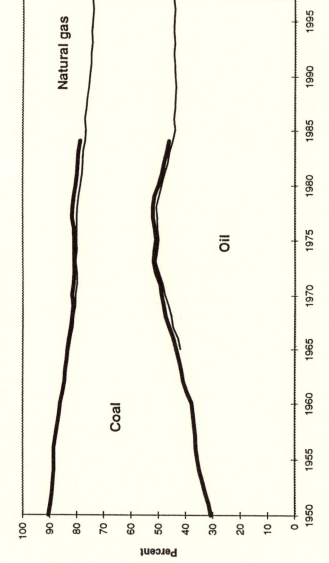

Sources: Statistics Norway, *North Sea Oil and Gas Yearbook* (1950–84, thick lines), and *BP Statistical Review of World Energy* (1965–97, thin lines).

Figure 2.2
World Consumption of Primary Energy, 1965–97

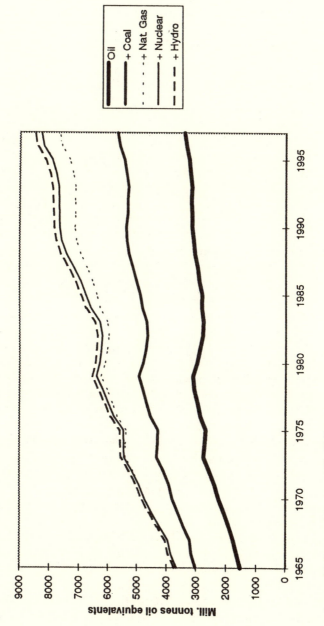

Source: BP Statistical Review of World Energy.

17

power besides coal, oil, and natural gas. Since the early 1970s nuclear energy
has become increasingly important, but its share is nevertheless no more than 7
percent (as of 1996). The contribution of hydroelectric power is even less, and
solar and wind power less still (not shown). The total demand for energy has
risen steadily, albeit with some "hiccups," from just under 4000 million tonnes
oil equivalents in 1965 to over 8000 in 1996.

SUBSTITUTION, PRICES, AND STRUCTURAL CHANGES

Unsurprisingly, prices seem to play a role in the choice between alternative
sources of energy. The huge rises in the price of oil apparently shifted the
demand for energy away from oil and to other sources. But some choices are
harder than others. Oil products are not easily displaced in certain types of use.
Oil is the only type of fuel used for aviation, and the alternatives for the auto-
mobile are expensive. It is instructive to look at what happened to the demand
for different oil products after the increase in oil prices in the 1970s. Figure 2.3
shows the shares of different oil products in total consumption since 1965. The
share of heavy fuel oil has fallen considerably, particularly after the second oil
price rise in 1979–80, whereas the shares of gasolines and in particular so-called
middle distillates have risen. Heavy fuel oil is the type of product most easily
replaced by other sources of energy. It is used for fueling electric power plants,
for example, where it can be replaced by natural gas, coal, or nuclear energy.
Note, however, that plants usually need to be converted or replaced by new ones
to accomplish this,[2] so the phasing out of oil is not to be expected to be an
immediate consequence of a rising price. First, the higher price must be expected
to be permanent, which will only happen if it has remained high for some time.
Second, it may well be economical to continue to use the old oil-fired equipment
for some time, even if it would be better to opt for some other fuel when the
equipment must be replaced. The fact that most of the drop in the demand for
heavy fuel oil took place after the second oil price rise could be either because
the second rise was taken as a sign that the price rise was indeed permanent
(which it was not), or that it was not economical to phase out old plants and
equipment running on oil until some years after the first oil price rise had made
investment in such plants and equipment unprofitable.

Gasolines include fuel used for motor cars, a product not easily replaced by
other fuels. Some "substitution" can be achieved, however, by forsaking comfort and speed and opting for cars that use less gasoline. The difference in design
and fuel consumption between European and American cars should drive home
the point that such substitution is to be expected to follow rising prices of
gasoline; the price of gasoline in the United States is only one-third to one-
fourth of what is common in Europe. Middle distillates include diesel oil but
also jet fuel, which has no substitute. The increase in aviation accounts for much
of the increase in the demand for middle distillates.

The oil price rises in the 1970s seem either to have coincided with or to have

Figure 2.3
World Consumption of Oil Products, 1965–97

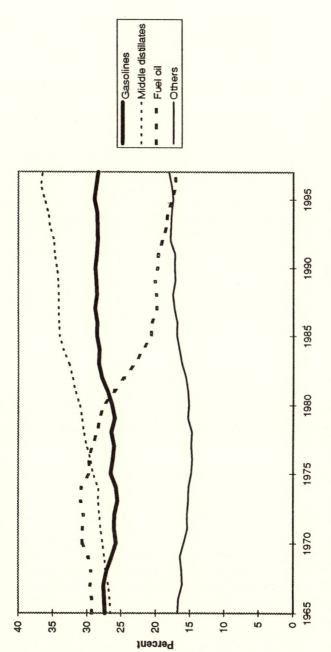

Note: "Gasolines" include aviation and motor gasolines and light distillate feedstock. "Middle distillates" consists of jet and heating kerosenes, and gas and diesel oils (including marine bunkers). "Fuel oil" includes marine bunkers. "Others" consists of refinery gas, LPGs, solvents, petroleum coke, lubricants, bitumen, wax, and refinery fuel and loss.

Source: BP Statistical Review of World Energy.

brought about fundamental changes in the relationship between economic growth and the use of energy in the most developed economies of the world. Figure 2.4 shows the annual growth rate of GDP and the use of energy over the three decades since 1960 in four areas of the world, North America, western Europe, the Pacific Region, and South Asia.[3] Although the growth in energy use exceeded the growth in GDP in the OECD countries in the decade 1960–70, it was considerably less in 1970–90. Whether this is due to the higher price of oil or to a structural effect accompanying increased affluence is not easy to tell. The economic growth in the OECD countries since the 1970s has mainly been in services, which are less energy intensive than industrial production, but whether this is due to a higher price of oil or to preferences for services rather than goods as incomes rise we do not know. Both factors probably play a role. In the countries of South Asia, all of which are poor, the growth in the use of energy has been almost as fast as the growth of GDP over the entire period. The picture from the Pacific region is more mixed, as is not surprising since this group of countries includes some that are very poor and some that are industrially highly developed. Figure 2.5 shows the GDP per capita for these four groups of countries, according to which the South Asian countries are clearly at the bottom of the heap. They, and the Pacific region, are also the ones with the highest and accelerating rate of energy growth. This indicates that the growth in energy use may be expected to be fastest in poor countries on their way toward overcoming poverty.

As a consequence of these changes, the OECD countries have lost much of their previous dominance in the energy markets of the world. Figure 2.6 shows the share of four categories of countries in the consumption of primary energy in 1973, on the eve of the first oil price hike, and in 1993. The share of the OECD countries fell from 65 percent in 1973 to 54 percent twenty years later. The share of the former Soviet Union and eastern Europe fell slightly, while the share of east Asia climbed from 8 to 14 percent.

THE WORLD TRADE IN OIL AND GAS

Table 2.1 shows the production and consumption of oil and natural gas by major areas of the world. This table reveals a basic difference between oil and gas, which has an important implication for world trade. In general there is substantial discrepancy between production and consumption of oil in all major areas of the world. This reflects the fact that oil deposits tend to be in countries and areas that are not major consumers of oil. The Middle East is the most important producing area for oil, producing almost one-third of the world total in 1996, but it consumes less than 20 percent of its own production. In Africa more than two-thirds of what is produced is exported out of the region. Other excess production areas are Central and South America and the former Soviet Union. The deficit areas are North America, Europe, and an area called Asia Pacific, which includes Australia, New Zealand, the Pacific islands, and all of

Figure 2.4
Annual Growth Rates in Gross Domestic Product (GDP) and the Use of Energy, 1960–90

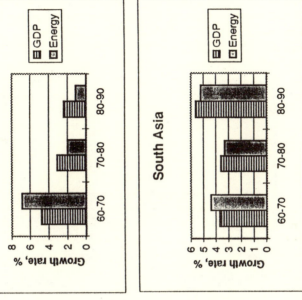

Source: World Energy Council, *Energy for Tomorrow's World* (London: Kogan Page, 1993).

Figure 2.5
GDP per Capita, in 1985 U.S.$

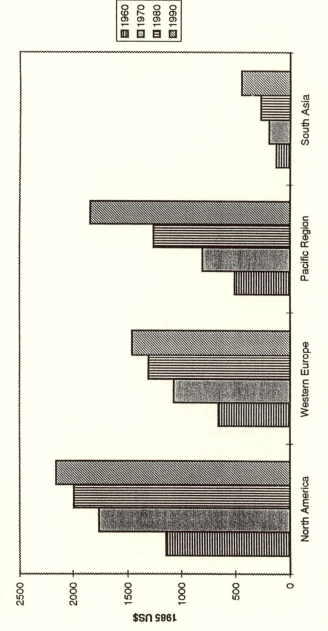

Note: North America and Western Europe include the OECD members of North America and Europe as of the early 1990s. The Pacific region is a very heterogeneous group, including Australia, the Pacific islands, Japan, China, and other East Asian countries. South Asia comprises Afghanistan, Bangladesh, Bhutan, India, the Maldives, Nepal, Pakistan, and Sri Lanka.

Source: World Energy Council, *Energy for Tomorrow's World* (London: Kogan Page, 1993).

Figure 2.6
Shares of Different Areas of the World in the Total Use of Primary Energy

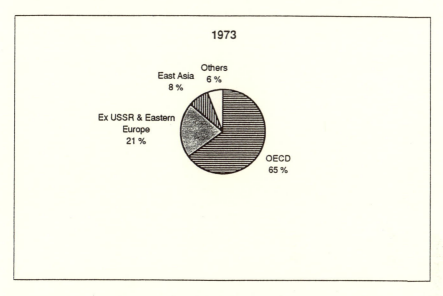

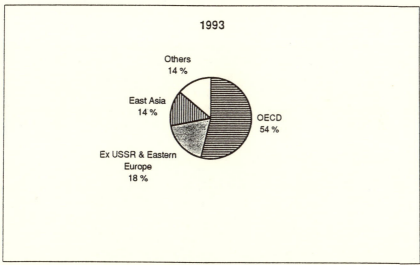

Source: BP Statistical Review of World Energy.

Asia except the Middle East. The countries of Asia Pacific consumed more than twice as much oil as they produced in 1996, and so did the countries of Europe. The main consumer of oil in Asia is Japan, which produces no oil of its own. In Europe only Norway and Great Britain produce oil in significant quantities

Table 2.1
Production and Consumption of Oil and Natural Gas by Area (1996)

	Oil (mill. tonnes)		Nat. gas (m.t. oil equivalents)	
	Prod.	Cons.	Prod.	Cons.
North America	660.7	986.3	658.0	663.7
South & Central America	313.9	203.7	75.7	75.5
Europe	328.1	740.1	250.8	376.4
Former Soviet Union	352.6	196.5	602.1	473.6
Middle East	983.3	190.5	135.2	128.3
Africa	359.6	110.3	82.2	43.1
Asia Pacific	363.5	885.4	204.7	211.0
World total	3361.6	3312.8	2008.7	1971.6

Source: BP Statistical Review of World Energy.

while the countries on the European continent are major consumers. The United States consumed more than twice its own production of oil in 1996.

The implication of this mismatch between the distribution of oil deposits and the location of major demand areas is that oil is a major commodity of international trade. A tremendous transportation and distribution industry has sprung up to bridge the gap between the regions of supply and demand. This mismatch has, furthermore, important implications for international politics, due to the importance of oil as a source of energy. The United States and its allies in western Europe would hardly have come to the defense of Kuwait after the Iraqi invasion in 1990 if the Arabian peninsula had been a province mainly of camels and Bedouins, but neither might the need have arisen.

Looking at the rightmost columns of Table 2.1 we see that the figures for production and consumption of natural gas match much better. Only in Europe, the former Soviet Union, and Africa is there a major difference between the production and consumption of natural gas. The reason is that natural gas is much more difficult to transport than oil. One thousand cubic meters of natural gas has roughly the same energy content as one tonne of oil, which takes up roughly one cubic meter of space. Hence gas requires about a thousand times more space than oil, at atmospheric pressure, for any given energy content. Furthermore, natural gas is much more difficult to handle than oil; it cannot just be poured like liquid, and if this were tried it would escape and dissipate into the atmosphere. Natural gas must therefore be compressed and possibly cooled to very low temperatures at which it becomes liquid. This is an expensive and cumbersome operation. Therefore, there is hardly any world market for natural

Figure 2.7
Prices of Natural Gas Imported to Japan (LNG), the EU (cif), the United States, and U.S. Wellhead Prices, 1975–97

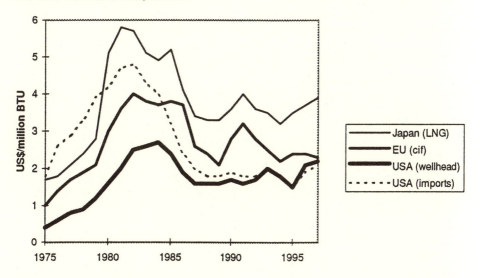

Source: *BP Statistical Review of World Energy.*

gas, as there is for oil, but mainly regional markets. Africa exports its surplus gas to western Europe, mainly through a pipeline from Algeria. In addition, Europe gets much of its gas supplies from Russia.

The absence of an international market in natural gas is well illustrated by the difference, and at times divergence, of the import prices of gas in the three regional markets of the United States, western Europe, and Japan (Figure 2.7). The price in Japan has almost consistently been the highest, but all gas imports to Japan consist of liquefied natural gas (LNG). The price of imports to the United States was very high in the 1970s but has since fallen, reflecting the virtual disappearance of LNG; virtually all natural gas that is imported to the United States is transported by pipeline from Canada. The imports to Europe also mostly come by pipeline, but the price is nevertheless considerably higher than the import price to the United States. Note, however, the similar long-term pattern of the three price series, due presumably to competition with internationally marketed substitutes such as oil and coal.

The importance of the Middle East becomes even more marked when we look at the shares of world reserves and world production, shown in Table 2.2. The Middle East produces 30 percent of all oil in the world but its share of world reserves is 65 percent, and its reserves would last almost a hundred years at the current rate of production. North America produces about 20 percent of the world total but its reserves are less than 10 percent. Despite the importance of oil in the Norwegian economy and Norway's national wealth, the North Sea reserves are minor on a world scale; the oil reserves in all of Europe are only 2 percent of the world total. It seems beyond doubt that the position of the countries in the Middle East will become more and more important in the future. This will strengthen their market power and possibly lead to a repetition of the price rises of the 1970s.

For gas the picture is somewhat different. The former Soviet Union, which largely means Russia, has 40 percent of the gas reserves of the world and produces almost one-third of the world total. The Middle East is next with 33 percent, but its share of world production is only about 7 percent, due to its small home market and long distances to major markets in Europe. North America produces slightly more gas than the former Soviet Union but has only 6 percent of world reserves. The gas reserves of Europe are somewhat more impressive than its oil reserves: 4 percent of the world total.

OIL AND NATURAL GAS

Crude oil is, as the name suggests, fairly useless in its natural state. To be useful it must be refined. Oil refining consists of separating the various petroleum products and removing impurities such as sulfur from the oil. Oil can be refined by heating it, as the different products evaporate at different temperatures, much as alcohol can be distilled from ignoble brews in one's own basement. The lighter the product, the lower is the temperature at which it evaporates. Natural gas evaporates at normal temperature under atmospheric pressure. Table 2.3 shows different categories of petroleum products and the temperatures at which they evaporate. Modern oil refining is, however, a lot more sophisticated than that, inter alia because the demand for light products has increased more than the demand for heavy products, making it necessary to produce more of light products from every barrel of crude. Untreated petroleum that is liquid at normal temperatures is what we call crude oil. Natural gas liquids (NGL) are gases that are liquid in the underground deposits because of the high pressure. Crude oil that contains a large proportion of molecules with few carbon atoms is referred to as "light."[4]

In contrast, natural gas does not need extensive treatment before being used. The difficulties associated with using natural gas lie in its bulkiness and the costliness of transportation, as already explained. In places far from markets for natural gas it has not been uncommon to burn the gas where oil and gas are found together, just to get out the oil. The uses of different oil products are indicated in Table 2.3. Natural gas is used for the same purposes except that it is hardly used at all in transportation and its use as feedstock for the chemical

Table 2.2
Share (%) of Production and Reserves of Oil and Gas in Major Areas and the Reserves/Production Ratio (R/P), 1996

	Oil			Gas		
	Prod.	Res.	R/P	Prod.	Res.	R/P
North America	19.7	8.3	18.1	32.8	6.1	11.8
South & Central America	9.3	7.6	36.1	3.6	4.2	70.2
Europe	9.8	2.0	8.2	12.5	3.9	18.6
Former Soviet Union	10.5	6.4	25.5	30.0	40.4	81.1
Middle East	29.2	65.2	93.1	6.8	32.5	>100
Africa	10.7	6.4	25.0	4.1	6.5	>100
Asia Pacific	10.8	4.1	15.7	10.2	6.4	40.1
World Total	100	100	42.2	100	100	62.2

Source: BP Statistical Review of World Energy.

industry is more common than for oil. Figure 2.8 shows the use of natural gas in the OECD countries in 1994. More than one-third was used for home heating, cooking, and in agriculture and service industries. Almost 30 percent was used by industry, some of it as chemical feedstock. About one-third was transformed, most of it into electricity and heat for further distribution to households and industries. Less that 10 percent was used in the oil and gas sector itself, and only 2 percent for transportation.

There was indeed a time when natural gas was thought of as too precious to be burned for production of secondary energy, because it could be burned directly or used as feedstock for the chemical industry. Both the United States and the European Union had regulations in place banning the further use of natural gas for the production of electricity. These regulations have now been abandoned, as natural gas is becoming increasingly attractive as a fuel for producing electricity. There are two reasons for this. First, natural gas is a much cleaner fuel than coal and oil. Emissions of soot and sulfur dioxide are negligible, and emissions of natrium oxide and carbon dioxide are much less than from coal and oil. The latter is extremely important, to the extent it is desired to cut down on emissions of greenhouse gases. Second, the conversion efficiency of burning natural gas for producing electricity has taken a quantum leap in recent years with the combined cycle turbine; the conversion efficiency has increased from 25–35 percent to more than 50 percent, and a turbine produced by Siemens is reported to have reached 58 percent conversion efficiency (Stoppard, 1996, p. 19).

Table 2.3
Boiling Points and Uses of Different Oil Products

Product	No. of carbon atoms	Boiling point	Main uses
Gas	1 - 4	-164 - +30	Fuel, carbon black
Petroleum ether	5 - 7	+30 - +90	Solvent, dry cleaning
Gasoline	5 - 12	+40 - +220	Motor fuel
Kerosene	12 - 16	+200 - +315	Stoves, diesel engines, rockets and jets
Gas oil (fuel oil)	15 - 18	up to +375	Furnaces, diesel engines
Lubricating oil	16 - 20	+350 and higher	Lubrication
Greases	18 and higher	Semisolid	Lubrication, sizing paper, medicines
Paraffin ("wax")	20 and higher	Melts at +51 - +55	Candles, waterproofing
Pitch and tar		Residue	Roofing, paving, protective paints, manufacture of rubber
Petroleum coke		Residue	Fuel, carbon electrodes

Source: Hussein K. Abdel-Aal and Robert Schmeltzler, *Petroleum Economics and Engineering.*

As already discussed, because of transportation costs there is hardly any world market for natural gas. Although natural gas can be transported in liquefied form in tankers just like oil, the trade in liquefied gas is of a limited scope, with Japan being the major recipient. In Europe and North America gas is overwhelmingly transported through pipelines. This makes for regional markets rather than a world market. The three biggest market areas are North America, the former Soviet Union, and Europe. Figure 2.9 shows the consumption of natural gas in Europe and the United States. The market in the United States

Figure 2.8
Uses of Natural Gas in the OECD Countries, 1994

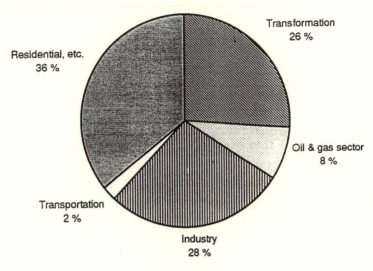

Transformation
26 %

Residential, etc.
36 %

Oil & gas sector
8 %

Transportation
2 %

Industry
28 %

Source: IEA (International Energy Agency), *Energy Statistics of OECD Countries 1993–94* (Paris: OECD).

comes across as mature, or saturated; the consumption of natural gas was no higher in 1996 than it was in 1971, and from 1972 it actually fell until the mid-1980s. The rise in consumption in the United States since the mid-1980s coincided with the deregulation of the market, to be discussed in the next chapter.

The market for gas in Europe was much later in developing. Europe is less well endowed with natural gas deposits than North America (see Table 2.2), and its deposits were discovered later. There are some indigenous gas fields in France, Italy, and Germany, but these countries are far from being self-sufficient in gas. It was the discovery of the very substantial Groningen field in the Netherlands and the gas reserves underneath the North Sea in the 1960s that gave the impetus to the use of natural gas in western Europe on a major scale. The consumption of natural gas in Europe increased rapidly in the 1960s and early 1970s; in western Europe it increased from only 6 billion cubic meters in 1960 to more than 200 in 1979 (Estrada, Moe, and Martinsen, 1995, p. 33). But after the first energy crisis (1973) the rate of growth became quite a bit lower, and consumption of natural gas actually fell from 1979 to 1983. Since 1983 the consumption of gas in Europe has grown steadily, and over the last few years (1994–96) the growth has been very strong.

The penetration of gas in the west European energy market has also risen. The share of gas in the consumption of energy rose from 6.3 percent in 1970 to 14.5 percent in 1980 and 17.3 percent in 1990, and is still rising. This not-

Figure 2.9
Consumption of Natural Gas in the United States and Europe (excluding the Former Soviet Union)

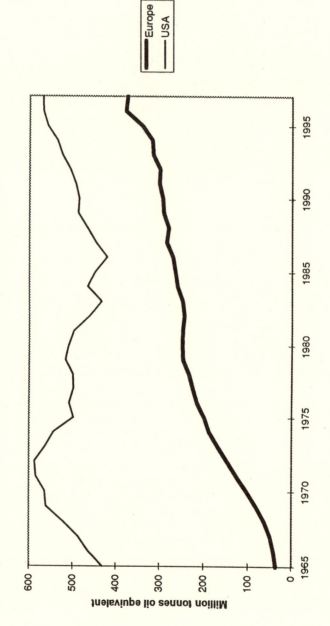

Source: BP Statistical Review of World Energy.

withstanding, it is often alleged that gas is being held back by lack of competition in the west European market (the market organization will be discussed in the next chapter). Even if the population and industrial development of western Europe is roughly comparable to the United States the consumption of gas is much less, as Figure 2.9 shows. The level of gas penetration in the different countries of western Europe varies considerably; in the Netherlands it is 74 percent, in the United Kingdom 54 percent, in France 25 percent, and in Spain only 5 percent (Estrada, Moe, and Martinsen, 1995, p. 325). The level of penetration is highest in countries with indigenous resource supplies, which is perhaps to be expected, but factors such as population density and geography also play a role. In Norway there is no domestic use of gas despite an abundant resource base. Low population density and mountainous terrain, together with abundant sources of hydroelectric power, make it uneconomic to distribute gas domestically through pipelines.

Another factor that may have slowed down the penetration of natural gas in western Europe is its dependence on sources outside the area. Western Europe gets a large share of its gas supplies from Russia and Algeria, with Nigeria, the Middle East, and Iran being potential suppliers. There is some fear that political upheavals might threaten the even flow of gas from these sources. Furthermore, the pipelines from Russia cross several state boundaries, which in fact have increased in number since the political upheavals in eastern Europe in the 1990s, and pipelines from the Middle East would also have to cross several international borders. This is a complicating factor since transit countries demand a fee for allowing pipelines to cross their territory, and political upheavals in the transit countries might also threaten supplies.

Nonetheless, the prospects for a further penetration of gas into the energy markets of Europe, and possibly elsewhere, appear bright. For one thing it is abundantly available (see Table 2.2). Second, the advantages of gas as a source for electricity have increased considerably in recent years. Investment in nuclear power has come to a standstill since the debacles of Three Mile Island and Chernobyl, because of the high risk perceived.[5] Furthermore, there are the environmental advantages over coal and oil and the increased efficiency of gas turbines, as already discussed.

ENERGY AND SUSTAINABILITY

Our ultimate source of energy is the sun, the Hydrogen Bomb Mother, as we might affectionately call her, for it is the radiation from the ongoing nuclear fusion process on the sun that fuels all life on earth. But we are not equipped to harness that energy directly; the plants have an edge over us in that respect. To utilize the solar energy it must first be converted into some, for us, useful form. We eat plants that have used sunlight to produce chemical substances that we can digest, or animals and fish that themselves live on plants or plankton.

We build shelters and make tools from wood, use wood or dung for fire to keep us warm or to fuel heat-driven production processes.

If someone finds these last examples out of date it is deliberate. The burning of wood and dung is something we associate with the past. But note that these processes, apart from emission of greenhouse gases, which we shall discuss in a moment, are what in popular parlance would be called sustainable, in the sense that they utilize the incoming flow of energy from the sun. The typical plant is grown over a season and so represents a stored-up energy of less than a year, domestic animals are raised and slaughtered periodically, but essentially all of this amounts to using the inflow of energy from the sun in a repetitive cycle.

The industrial revolution changed this in one fundamental respect. It made accessible to us new forms of energy that multiplied our productive capacity and increased our comfort to an extent our ancestors could hardly dream of. In fact, large parts of humanity still have very limited access to the fruits of the industrial revolution. But the forms of energy harnessed by the industrial revolution represent stored-up sunshine from past millennia. Coal, the king of the industrial revolution, is fossilized vegetation; petroleum, the upstart, is fossilized marine organisms. By burning petroleum and coal we are using up the gifts of the past. This is not sustainable in the sense that it can be carried on forever. But no one knows how much coal and oil is in the ground. Not even the utilization of sunshine is sustainable in the above sense; there will come a time when the sun burns out, but that will not happen within a time frame that we need worry much about.

At some time, or over time, we may have to make the transition from fossil fuels to renewable energy, renewable, that is, in the sense that it is based on the continuous flow of sunshine and not a depletion of stored-up sunshine harnessed by organisms that lived millions of years ago. There are processes such as photovoltaic cells that do this by converting the flow of sunshine into electricity. Other processes, such as windmills, hydroelectric power plants, and tidal power plants, utilize forces that are driven by sunshine (climatic phenomena like wind and rainfall) or gravity (tidal waves, hydroelectric power). The technology to use these processes is already here, but they are either expensive, environmentally offensive, or not very abundant in an easily accessible form. As fossil fuels become more scarce and expensive these sustainable energy forms will become more advantageous and they will increasingly replace the fossil fuels. Nuclear fusion is another undepletable source of energy but the technology to master it has still not been harnessed.

For a long time to come the fossil fuels will reign supreme. As the twentieth century draws to a close they are responsible for about 90 percent of all commercial energy produced in the world. A recent investigation by the World Energy Council into the likely future pattern of energy use concludes that the total amount of fossil energy used will continue to rise well into the twenty-first century, as the total amount of energy demanded will continue to rise even if the percentage supplied from fossil forms may decrease slightly.[6]

This prediction raises another issue of sustainability. The increased concentration of carbon dioxide (CO_2) and other greenhouse gases is believed to be causing an increase in the global mean temperature, a process that would go on for decades even if the emission of these gases were to be stabilized or even cut back. There is tremendous uncertainty about the magnitude of these changes and no less uncertainty about the consequences thereof. However that may be, if it is desired to avoid or to reduce these changes the emissions of greenhouse gases will have to be reduced. Since fossil fuels are responsible for a substantial part of the man-made greenhouse gas emissions, a reduction in the use of these forms of energy is a method that quickly comes to mind.

Such reduction will, however, be easier said than done. As already mentioned, the use of fossil fuels is expected to increase and not to decrease for many years to come. Substituting natural gas for oil or coal may alleviate the situation somewhat, but the emissions of carbon dioxide are nevertheless likely to continue rising well into the twenty-first century. There are ample reasons to believe so. Much of the increase in the use of energy is expected to happen in the economically less developed countries. If the people of these countries are to have a standard of living remotely resembling that which the inhabitants of the richest countries in the world take for granted they will need to increase their use of commercial energy many times over. This is necessary not just for communication and industrial processes, but also for relieving people from physical toil and unsanitary conditions such as breathing the thick smoke of cow dung from indoor fireplaces. To put things in perspective, the use of primary energy per capita in 1990 was 7.82 tonnes oil equivalents in North America and 3.22 in Western Europe, but 0.58 in sub-Saharan Africa, 0.76 in China and other Communist countries of Asia, and 0.52 in South Asia. At the same time, the population increase is largest in what euphemistically are called the developing countries. From 1960 to 1990 these countries increased their share of the world population from 68 to 76 percent, and this percentage is expected to rise further to 85 by the year 2020.[7]

Hence, if the emissions of carbon dioxide are to be stabilized or cut back at least one of two things must happen. Either the poor masses of the world will continue their toil in poverty or the inhabitants of the rich countries will have to cut back their standards of living to levels few would be willing to contemplate. Neither is likely to happen. The poor masses of the world will not with equanimity continue to live in squalor to preserve the comfort of those who presently are doing better and whose standards are widely broadcast by modern television, probably appearing more enviable than they are in reality. Rich people, on the other hand, have never been known to give up their privileges gladly. Indeed, in front of this stark choice the question must be asked if the cure is not worse than the problem. After all there is tremendous uncertainty surrounding the potential effects of increased concentration of greenhouse gases,[8] but any change in the climate will come gradually and provide some time for adjustment. A massive transfer to nuclear power would go some way toward squaring this circle, but that method has its own problems.[9]

NOTES

1. The renowned English economist Stanley Jevons wrote a book entitled *The Coal Question* on this issue, published in 1865, and calculated that Britain would run out of coal by 1960 at the present rate of extraction. The Swedish economist David Davidson also addressed this question in the journal *Ekonomisk Tidsskrift* in 1900. Maybe it is because of this experience that contemporary economists have remained aloof with respect to questions of imminent resource scarcity.

2. Some power plants can be run on alternative fuels (coal and fuel oil, for example) and thus have a built-in capacity for immediate response to price changes. It appears to have become more common to construct plants in this way after the oil price rises in the 1970s.

3. These figures are taken from *Energy for Tomorrow's World*, a report from the World Energy Council (London: Kogan Page, 1993). North America and western Europe include the OECD members of North America and Europe as of the early 1990s. The Pacific region is a very heterogeneous group, including Australia, the Pacific islands, Japan, China, and other East Asian countries. South Asia comprises Afghanistan, Bangladesh, Bhutan, India, the Maldives, Nepal, Pakistan, and Sri Lanka.

4. Note the two ways of measuring oil, in volume and weight. The commonly used "barrel" is a volume unit, corresponding to 0.159 cubic meters, or kiloliters. Due to differences in composition, oil from different fields has different specific weight, so one barrel does not weigh the same irrespective of where it comes from. North Sea oil is of the light variety but there are differences even there, one liter of oil from Ekofisk weighs 0.827 kilogram while one liter of oil from Gullfaks weighs 0.864 kilogram.

5. Whether this risk is just perceived or objectively assessed is unimportant, as long as it has an overriding influence on decision making. The human attitude toward risk is a fascinating subject; the general public seldom thinks in terms of probabilistic calculations and is much impressed by spectacular but rare events. More people are afraid of flying in a large, commercial jet than of driving the family car, even if the risk of a fatal accident in the latter is far greater. An interesting case of precaution occurred in Sweden in the 1970s. A satellite was expected to fall down somewhere in the wilderness of Lapland. One single person was living within the area that might conceivably be hit by the satellite debris. A helicopter was sent, under hazardous weather conditions, to fly him to a safer place.

6. World Energy Council, *Energy for Tomorrow's World*, (London: Kogan Page, 1993).

7. The figures in this paragraph are taken from ibid.

8. It may be worthwhile to point out that without the greenhouse effect we would not even exist, as the global mean temperature would be more like -18 degrees centigrade rather than $+15$, but there can be too much of a good thing.

9. In fact, throughout human history there have been changes in climate that people with more primitive technology have adjusted to. Some, however, apparently did not

succeed. The Icelandic settlement in Greenland disappeared in the Middle Ages, possibly because of a colder climate (the so-called Little Ice Age), and Iceland itself was almost vacated for the same reason. In fact, over several hundred thousand years ice ages have been the norm; the last 10,000 years or so have been abnormally warm. Perhaps a stronger greenhouse effect will only succeed in averting a new ice age, which some believe is otherwise overdue.

Chapter 3

Natural Gas

CHARACTERISTICS OF NATURAL GAS

Natural gas is, like oil, a form of petroleum produced by chemical reactions underground in the remains of organic material from earlier ages.[1] But there is one major difference. As discussed in Chapter 2, natural gas is bulky, difficult to handle, and expensive to transport. There are two ways of transporting natural gas, by pipelines or in tanks in liquefied form. In a pipeline, gas flows from one end where the pressure is high to the other end where it is lower. For pipelines that traverse long distances it is necessary to boost the pressure at several points to enhance the flow. For transportation in liquefied form the gas has to be cooled to temperatures of -160 degrees centigrade in special plants, pumped into tanks where it is kept cool during transportation, and pumped into storage at the other end of the journey. This is commonly referred to as liquefied natural gas (LNG). To make it usable, it must then be regasified and distributed to the end users through pipelines.

The economics of these transportation processes are roughly as follows. Liquefaction and regasification plants are expensive installations, but the cost of transporting LNG is comparatively low and proportionate to distance. The cost of a pipeline is roughly proportionate to its length, for any given dimension of the pipeline. The non-distance-related costs of the LNG system are high enough to make that process uneconomic vis-à-vis the pipeline method up to a certain distance, but as the fixed costs of the LNG system are spread over longer distances it eventually outperforms pipelines. Figure 3.1 shows representative costs of transporting gas through pipelines versus in liquefied form. LNG typically outperforms offshore pipelines at distances over 1500 kilometers, and onshore pipelines at distances over 3500 kilometers. This is exemplified by the fact that

Figure 3.1
Representative Costs of Oil and Gas Transportation, in U.S.$ per Million British Thermal Units

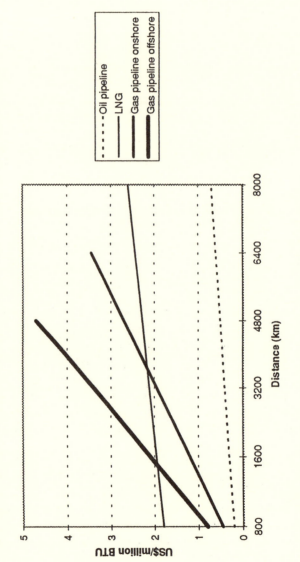

Source: OECD/IEA (1994), *Natural Gas Transportation*, after Jensen Associates, Inc.

gas is transported overland in pipelines in North America and Europe, and across the North Sea and the Mediterranean, whereas it is transported in liquefied form from Africa to North America, and from Indonesia, Australia, and the Middle East to Japan.

The low energy-to-volume ratio for gas makes heavy demand on storage space, to cope with seasonal or even diurnal variations in demand. Much of natural gas is used for generating electricity, space heating, and cooking. The use pattern in these applications is highly seasonal and dependent on weather and climate, especially for space heating and air conditioning. This leads to corresponding seasonal variations in the demand for natural gas.[2] Some storage can be obtained in the pipelines by increasing the pressure, but usually much more storage capacity is required than can be provided in this way. Owing to the volume requirements, land-based tanks are not a solution. Underground storages such as disused gas fields or mines, aquifers, or washed-out salt domes are typically used for coping with demand fluctuations. The provision of storage capacity is therefore an important part of the services of gas distribution and pipeline companies.

The high capital costs of transporting gas have two implications for industrial structure. First, gas producers have been reluctant to develop gas fields unless having secured a long-term commitment from buyers and, quite often, a joint venture in building the necessary transportation systems. Typically the gas sales contracts have included so-called take-or-pay clauses; i.e., the buyer has been obliged to pay for a certain quantity of gas whether he was prepared to receive it or not. This ensured a smooth financial flow to defray the heavy capital costs of gas field development and pipeline construction. Often, however, the buyers could offset their payments against receiving gas at a later date in excess of the minimum they contracted for. This differs radically from the marketing of crude oil. The storage and transportation system for oil is uncomplicated enough that there is no problem to get rid of whatever one produces at a price the market will bear, or to obtain any quantity one might need (except in highly special circumstances such as the Arab oil embargo in 1973, and even that was not particularly effective).

Second, the high capital costs of pipelines and LNG systems make them a classic natural monopoly case. Most of the cost of transporting gas in a pipeline is fixed capital cost; the cost of increasing the flow of gas is comparatively small. Figures from the United States classify over 90 percent of the cost of transporting gas as fixed capital costs (OECD, 1994, p. 158). This natural monopoly character is enhanced by the fact that pipelines display increasing returns up to a point; pipeline costs are roughly proportionate to the radius (or circumference) of the pipeline while the transportation capacity is roughly proportionate to the radius cubed.[3] A wider pipeline will, however, have to have thicker walls. This and other inconveniences associated with bulkiness put a limit on how wide pipelines can be without becoming uneconomic or impossible to construct. The largest pipelines measure about one meter in diameter.

Figure 3.2
Equilibrium Prices (*P*) and Quantities (*Q*) Transacted of a Commodity under
Monopoly (*m*), Full Cost Recovery (*fc*), and Perfect Competition (*c*)

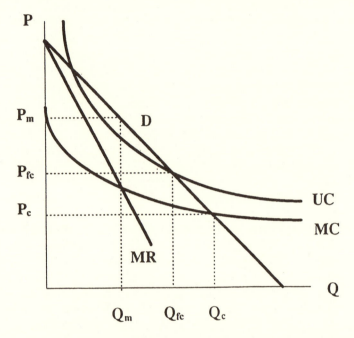

Note: A monopolist sets a price where marginal revenue (*MR*) is equal to the marginal cost (*MC*),
    full cost recovery requires that the price be equal to the unit cost (*UC*), while economic
    efficiency implies that the marginal willingness to pay, given by the demand curve (*D*), is
    equal to the marginal cost (*MC*).

THE GAS MARKET AS A NATURAL MONOPOLY

The natural monopoly poses a well-known economic problem, illustrated in
Figure 3.2. The demand curve D shows the willingness of the buyers to pay for
an additional unit of gas. The cost of transporting an additional unit of gas is
low, given that there is sufficient capacity in the pipeline. The curve MC shows
the cost of producing and transporting one additional unit of gas. The capital
costs are high, and the total cost per unit transported is well above the marginal
cost (MC), as shown by the unit cost curve (UC). To make ends meet with a
uniform tariff, the tariff has to exceed the marginal cost. This, however, will
discourage some who are willing to pay what it costs to produce and transport
an additional unit of gas from buying that gas; raising the tariff from the marginal cost level (P_c = MC) to the full cost coverage level (P_{fc}) will reduce the
demand from Q_c to Q_{fc}. But the monopoly will opt for a still less desirable

solution. If it applies a uniform tariff and knows the demand schedule it will recognize that it has to lower its tariff to attract an increased volume of transportation. The marginal revenue for the transporting firm is therefore less than the buyers' willingness to pay for an additional unit transported, as shown by the marginal revenue curve (MR) in the figure. The profit-maximizing volume for the natural monopoly is where the marginal revenue is equal to its marginal cost (MR = MC). This implies a still higher tariff (P_m) than the full cost solution and less volume transported (Q_m instead of Q_{fc}), and excludes still more units for which the buyers are willing to pay the additional transportation cost.

One possible and practical solution to this problem is to have the monopoly apply a two-part tariff where one part is payment for access to transportation capacity and the other is payment for the use of the capacity. In this example the buyers of gas would have to pay a relatively high amount for the privilege of using the pipeline and a much lower amount, equal to the marginal cost, for the actual offtake of gas. This tariff structure could in principle accomplish the twin goal of recovering the capital costs without discouraging the transportation of gas for which the buyers are willing to pay the marginal cost. In practice there are problems of calculating or even defining the true marginal cost (see OECD, 1994, Ch. VI). This is the principle that since 1992 underlies the regulation of the American natural gas pipeline industry.

The difference between oil and natural gas can be compared to the difference between road and rail transportation. Like natural gas pipelines, railroads are inflexible; they require substantial initial investment. Usually a railroad company has monopoly on owning and operating the railroad, and access to it is at the discretion of the company that controls the signal system, the stations and yards, and so forth. Like the transportation system for oil, ordinary roads are relatively flexible; while the most elaborate motorways constructed to take the heaviest traffic flows are expensive and elaborate, the country road is simple and inexpensive and yet accessible to the same equipment as the motorway; access to roads is easy even if it can be made discretionary by installing the necessary barriers and gates, such as is done for toll collection on motorways. Anyone who wants to set up a transportation company can have access to the road by simply paying the required toll if there is any.

Not surprisingly, the ownership of railroads was much fought over in the heyday of this mode of transportation and gave rise to many of the spectacular fortunes gathered in the late nineteenth- and early twentieth-century United States. Tariff practices were contentious, and the railroads eventually came under government supervision before largely becoming technologically obsolete. One way of regulating railroads, although not much in use anywhere, would be to require the railroad companies to allow anyone's rolling stock access to the railroad. This, presumably, would provide competition in transportation and lower tariffs. This option has come into vogue with respect to natural gas pipelines in North America over the last decade or so and is known as third party access.[4] Recently it has been introduced in the United Kingdom as well, and the European Union (EU) is moving toward this kind of solution. The controversial "Gas Directive," agreed upon by the EU energy ministers in December 1997, partly opens up the European natural gas market to competition. Large industrial buyers will be able to shop around for deliveries and to have the gas transported through the existing pipelines.

The theory of natural monopoly has been one of the underpinnings of the regulation of the natural gas industry in Europe and the United States. Economies of scale in pipeline transportation would either make competition impossible or lead to wasteful competition in the absence of regulation but, since a monopolist is likely to charge a price that is higher than the optimal one, gas prices need to be regulated. Suppose there is a natural gas monopoly charging the monopoly price P_m in Figure 3.2. Another company might be attracted by the high monopoly profit and enter the market, with the result that the price would fall and the unit cost for both firms would be higher than the unit cost at Q_m, because of unused capacity by the incumbent firm and underutilized economies of scale by the entering firm. Both might in fact end up making a loss, in case demand turns out to be less elastic than predicted by the entering firm. Preventing wasteful competition appears to have been an important consideration in the regulation of the interstate pipelines in the United States. The construction of new pipelines was regulated by the federal government, and pipeline companies had to convince the federal authority in charge of the natural gas industry of the need for the new line, and be able to show that it had a sufficient supply of gas for twenty (later twelve) years.

The regulation of the American natural gas industry has been severely criticized by a number of authors (Bradley, 1995; DeVany and Walls, 1995). The construction of natural gas pipelines crossing state boundaries (interstate pipelines) and the rates charged for their services were subjected to regulation by the federal government in 1938, and the wellhead price of gas going into such pipelines soon followed. The regulatory regime has undergone several changes and the interstate markets are now largely deregulated. It is, to say the least, open to doubt how well the regulation has worked over the years. The history of gas regulation, like any other government regulation, is a strong reminder that market failures cannot be corrected without cost; there is no such thing as a perfect and ideal regulatory regime. Government regulations are subject to their own imperfections, just as real-life markets are. Most importantly, government regulations are driven by political expediency and ambitions of government officials instead of profits. It is by no means clear that government regulations outperform imperfect markets in serving the interests of the general public.

It is possible to distinguish between two types of cost of regulation. One has to do with cost-enhancing incentives. When rules and procedures are substituted for the market mechanism as a driving force, firms have an incentive to incur costs to comply with rules and procedures in a way that will enhance their profits. Such procedures need not benefit the final consumer in any way, so the cost will be a pure waste. A story is told (Bradley, 1995, p. 438) of a drilling rig being moved to a hilltop to take advantage of higher regulated prices of gas from wells deeper than 15,000 feet. The second type of cost is the cost of enforcing the rules and regulations. Doing so requires an army of government officials, inspectors, and lawyers, who otherwise could be devoting their skills to other and often more productive purposes. The American Bar Association has good reason to be grateful for the regulation of the American natural gas industry.

For more than ten years there has been a trend both in the United States and elsewhere to deregulate markets and instead correct market imperfections by creating incentive structures that emulate the perfect market of the economics textbook as closely as possible. The mechanism that has been tried for natural gas is third party access, to which we now turn.

THIRD PARTY ACCESS

What is third party access and what are its implications? Third party access means that a pipeline company must allow any buyer and seller of gas access to its pipeline at a reasonable tariff, provided there is capacity available. This, needless to say, raises many questions. Who will be given priority if the demand for transportation exceeds the capacity to provide it? And what is a reasonable tariff? Traditionally pipeline companies sold a "bundle" of services to their customers; transportation, storage (to deal with peak versus offpeak demand), and the gas itself. Some were even producers of gas as well. British Gas, for example, had interests along the entire chain, from the wellhead to the burner tip, and had a monopoly of distributing gas in the whole of the United Kingdom. Its costs statements therefore provided limited guidance as to what was the true cost of transporting gas between any two points. Hence, in countries like the United States and the United Kingdom where gas markets have been deregulated, it has been required that the services of pipeline companies be "unbundled" to make clear what costs are due to transportation, storage, and the gas itself. Furthermore, transportation tariffs and schedules must be announced. British Gas was required to separate its transportation and marketing services and has now been broken up into five companies.

This notwithstanding, the principle is clear enough. Any two parties interested in moving natural gas from point A to point B should be allowed to use a pipeline that happens to connect these two points, even if it is owned by yet another party. The consequences would also appear to be clear enough. Competition must increase; the ability of the pipeline company to restrict the flow of gas and to maintain high tariffs would become less. The total supply of gas should increase, and the price of gas, and not just the transportation tariff (which is anyway a high portion of the final price), should come down.

Third party access has been introduced in the United States and Great Britain and has had a profound impact on the natural gas markets in these countries. In the United States third party access started already in the 1980s. At this time an oversupply, the so-called gas bubble, developed, partly because of deregulation of wellhead prices, which increased supply, and partly because oil prices started to slip, making gas less competitive with oil. Some pipeline companies found themselves with gas that they had to take, because of their take-or-pay contracts, but could not sell. Some of them welcomed the possibility to be relieved of these obligations against transporting gas that producers sold directly to buyers. In the United States it appears that third party access was well received by the pipeline companies, which could and did apply to the Federal Energy Regulatory Commission for status as open access pipelines. Most big pipeline companies in the United States are now open access pipelines (DeVany and Walls, 1995).

Open access works in the United States as follows (DeVany and Walls, 1995). The previous customers of the pipeline companies now hold rights to transportation capacity for which they pay a tariff; in addition, they pay a tariff for the quantity of gas actually transferred (this is a variant of the two-part tariff discussed above). Once a month the holders of the rights to capacity give notice of how much of that capacity they intend to use the coming month. They can even transfer their rights temporarily to third parties. The unused capacity is then put up for hire in the spot market. Spot markets for gas also developed in the wake of third party access, and since 1990 there has been a futures market for natural gas on the New York Mercantile Exchange.

The emergence of the spot markets has been facilitated by the interconnection of the American pipeline grid. Several hubs exist where major pipelines meet and smaller lines fan out to various local markets. This, and the fact that there is third party access, makes it possible to sell gas by displacement; that is, gas can be sold to a distant place without the gas molecules moving that far; the entire load of the line is displaced, as it were, toward the buyer. In this way gas has been sold from Alberta (Canada) to Florida, for example. This interconnectedness has resulted in gas prices following a similar pattern in different locations.

The deregulation of the natural gas market in Great Britain is still underway and was expected to be complete by 1998. In some respects it may even be said to be more far reaching than in the United States; third party access in the United States does not apply to the local distribution companies; gas must be bought at the "city gate," except for large industrial buyers, some of whom are in a position to buy directly from the trunk lines. In 1996 households in a small "trial" area in the south of England were allowed to choose their own suppliers of gas. About sixty different suppliers took up competition for customers, offering bargain prices, and the market share of British Gas fell dramatically. The intention is to make similar arrangements for all areas in Great Britain.

The deregulation, in both the United States and the United Kingdom, has resulted in lower prices, as intended. Was the regulated monopoly solution, then, a mistake? It is noteworthy that both the United States and the United Kingdom, particularly the former, are mature markets, there exists an interconnected grid of pipelines, there are many suppliers (in the United States almost 30,000 [Jensen, 1992]), and a large infrastructure is in place. Would these investments have taken place in the beginning without the safeguards that exclusive rights and long-term contracts provided? We do not know. It also remains to be seen how well the existing distribution grid will be maintained and extended under the new regime. Authors like DeVany and Walls (1995) argue, vigorously and persuasively, for deregulation. One particularly persuasive argument they put forward is how decentralized markets economize on information and provide flexibility of response; rather than focusing on the static efficiency of markets, they emphasize the evolutionary character of markets and the elimination of inferior solutions. On the other hand, they hardly deal with long-term issues of establishing and maintaining the infrastructure. Futures markets are no substitute for long-term contracts, as they extend only a few months ahead. Perhaps the most fruitful way to look at deregulation of the gas market is to view it as a stage in a maturing process; it will work in a market of many players where the infrastructure is already in place and, hopefully, the infrastructure will be maintained and developed on an incremental basis. One thing worthy of note is that deregulation is in fact a bit of a misnomer; there is a need for a regulatory authority to make sure that access is in fact open and that pipelines will not revert to becoming local monopolies in case there are only a few players at both ends.

In continental Europe there has been skepticism and outright hostility to deregulation of the gas market, first proposed by the European Commission in 1992. In all countries on the continent the market for gas is either monopolized by law or characterized by a high degree of concentration (see Table 3.1). The pipeline companies on the continent have been unanimously and adamantly opposed to third party access. It is tempting to dismiss this as a reflex triggered by the fear of losing one's privileges. Still, we should not dismiss the issue so lightly. The natural gas markets in the United States and the United Kingdom are quite different from the markets in continental Europe. The market in the Untied States is characterized by many buyers and sellers and the country is crisscrossed by pipelines with a variety of owners; often a buyer has a choice of several sellers, and vice versa, and there is, as already stated, an organized spot market in natural gas. The market in the United Kingdom is still isolated from the continent, although this will soon change with the Interconnector pipeline being built from Bacton in the United Kingdom to Zeebrugge in Belgium. In the United Kingdom a large number of gas producers compete with one another in areas that have been opened up for competition.

In continental Europe the situation is different. Not only is the transportation of gas in each country typically handled by national monopolies (cf. Table 3.1), on the supply side there are only a few players as well. The Russian gas being piped into Europe is controlled by one company (Gazprom). The Norwegian gas is sold through a national committee of gas producers (*Gassforhandling-*

Table 3.1
An Overview of the Natural Gas Market in Continental Europe and Its Main Features of Organization

France

Gaz de France has monopoly rights to import and distribute gas to local distributors. Ninety percent of all gas is imported, and most local distribution is in the hands of Gaz de France.

Belgium

Distrigaz has monopoly rights to distribute gas to local utilities. All gas is imported.

The Netherlands

Gasunie has a virtual monopoly in the distribution to local distribution companies. The Netherlands is a major gas producer and exporter but nevertheless imports a small amount (about 5 percent).

Germany

There are a number of pipeline transmission companies, which have an agreement about areas of operation. Ruhrgas is the largest, covering about 70 percent of sales. The newcomer Wingas, jointly owned by the Russian gas company Gazprom and the German chemical giant BASF, appears to be an increasing threat to incumbent companies in the German market. Germany produces a significant amount (20 to 30 percent) of the gas it uses.

Italy

The state-owned company SNAM has a virtual monopoly on imports and transmission of gas to local distributors.

Source: Based in part on Estrada, Moe, and Martinsen (1995).

sutvalget), and the Algerian gas is sold by the national Algerian company Sonatrach. The main internal producer in the European Union, the Netherlands, sells its gas through the national company Gasunie. While the European Union could deregulate its own production if it so desired, and possibly the Norwegian production as well through the European Economic Area Treaty, it would have no power over Russian and Algerian or other foreign supplies. There are only a few pipelines controlled by the major producers through which the gas flows from Russia, Norway, or Algeria into Europe. It is not obvious that third party access would create a lot of competition and lower prices under those circumstances.

In the following two sections we shall consider formal models with only one or two suppliers. The conclusion that emerges from these models is that deregulation of the pipeline companies need not result in a greater benefit for the final customers but only in a transfer of monopoly profits from the pipeline companies to the suppliers. The suppliers might find it in their interest to restrict supplies rather than compete, and without increases in supply there will be no fall in prices for the final consumer. These conclusions may, however, be too pessimistic for continental Europe; there are more than just two suppliers, and the likelihood of competition increases with the number of suppliers. These simple models of one or two suppliers are nevertheless of some methodological interest, as they demonstrate how a market with only a few participants on the supply side may be analyzed, but the analysis becomes progressively more complicated as the number of suppliers is increased.

How likely is it that the continental gas market will be deregulated? What are the driving forces behind deregulation? In the United Kingdom, and possibly in the United States as well, the deregulation process seems to have been driven by ideology, i.e., a perception of benefits of competition among politicians and civil servants, rather than by pressure from buyers and producers of gas. On the continent this ideology is mainly represented in the European Commission, which in 1992 proposed deregulation, while the pipeline companies and most national governments in the EU have been hostile. Possibly large buyers, such as electricity producers and industrial companies, will exert pressure for deregulation, in the hope of getting increased supplies at lower prices. The German chemical company BASF appears to be heading for greater competition through its subsidiary Wingas.

FORMAL MODELS OF THIRD PARTY ACCESS

Suppose a producer delivers to a gas transmission company in another country, which in turn sells the gas to many local distribution companies. This is not an unrealistic representation of the European continental scene as it has been up to now. Presumably the pipeline companies are faced with a downward sloping demand schedule for gas to be sold to the local distribution companies, as shown in Figure 3.3.[5] If the pipeline company buys the gas from the producer at a given price S and sells it to the distribution companies at a price P that depends on the total quantity sold (Q), the profit of the pipeline company (π_t) will be

$$\pi_t = P(Q)Q - SQ - C_t Q, \tag{3.1}$$

where C_t is the transportation cost per unit of gas. For simplicity, we assume a fixed unit cost of transportation, even if there are substantial economies of scale in this activity, which is why this activity tends to be monopolized and highly concentrated. We may think of C_t as the unit cost in a pipeline of an optimal size, to be determined simultaneously with finding the optimum quantity and price.

Maximization of the pipeline company's profit entails

Figure 3.3
Nash Bargaining Equilibrium in a Market for Gas with a Pipeline Company as the Single Buyer and One Producer as the Single Seller (Bilateral Monopoly)

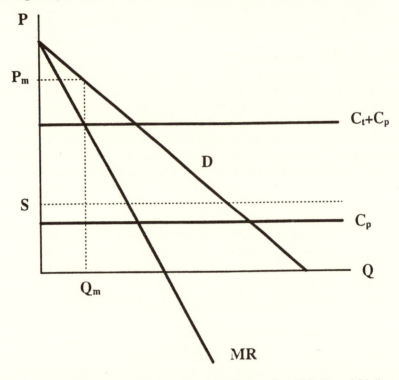

Note: Q_m is the quantity that maximizes the total profit. Market price will be P_m, and S is the price the supplier gets, which divides the profit equally between the buyer and the seller. C_p is the unit cost of production and C_t the unit cost of transportation.

$$P + P'Q \ (= MR) = S + C_t; \qquad\qquad (3.2)$$

that is, the company would like to sell the quantity where the marginal revenue (*MR*) is equal to its marginal cost of gas, which is the sum of the unit transportation cost and the price to be paid to the producer for obtaining the gas.[6] But the price of the gas is not independent of the quantity the pipeline decides to buy from the producer. The pipeline company will have to bargain with the producer over the price it pays and the quantity it sells, and the producer will consider what he gets from any possible constellation of price and quantity. With a given unit cost C_p of producing the gas, the profit of the producer (π_p) will be

$$\pi_p = (S - C_p)Q. \qquad\qquad (3.3)$$

Each party will want to maximize its own profit, but there is no single solution that maximizes each party's profit simultaneously; there is a cake to be divided, and one party's share will be at the other's expense.

There is an infinity of possible outcomes to such "cake-dividing" problems. One popular method of picking a solution is to assume that both parties are interested in making the cake as large as possible and dividing it equitably.[7] The sum of profits is

$$\pi_t + \pi_p = P(Q)Q - SQ - C_tQ + (S - C_p)Q = P(Q)Q - (C_t + C_p)Q. \quad (3.4)$$

The only role of S, the price at which the pipeline company buys gas from the producer, is to divide the profit between the two parties; the total profit depends only on the quantity produced and distributed. Maximizing the sum of the profits with respect to Q entails

$$P + P'Q \ (=MR) = C_t + C_p. \quad (3.5)$$

This is identical to the optimum solution for the pipeline company as a monopolist (the point P_m, Q_m in Figure 3.3). The difference is the price the pipeline company pays to the producer. This is the parameter that divides the spoils between the producer and the pipeline company. The equitable solution is the one where both get the same profit, i.e., $\pi_t = \pi_p$, which entails

$$S = C_p + \frac{1}{2}(P - C_t - C_p); \quad (3.6)$$

that is, the producer gets a price that covers his unit cost of production and gives him one-half of the profit per unit.

How would third party access affect this solution? In this world of bilateral monopoly, it would mean that the pipeline company would be obliged to allow the producer access to its pipeline at a tariff that would only allow the pipeline company normal return on its capital. The producer would be able to get the entire profit, so his net price would become $P - C_p - C_t$. He would, however, still be free to limit his sales to whatever would maximize his profits, and he would go for the monopoly quantity but now obtain a higher price.

The effect of third party access in a world of a bilateral monopoly between a pipeline company and a gas producer would thus be to concentrate the monopoly power in the hands of the producer; the final buyers of gas would have little to gain. The only way in which the final buyers would gain would be through intensified competition among different producers, with producers taking advantage of having access to the pipelines at reasonable tariffs and increasing the total supply and driving down the price of gas.

That third party access would not provide much benefit to consumers in a world of only one supplier is fairly evident, but would it happen in the real world of continental Europe with three external suppliers, Norway, Russia, and Algeria, and indigenous suppliers such as the Netherlands and, to a lesser extent, Germany, and possibly further external sources such as Great Britain, Nigeria, and the Middle East? That question is not easily answered, but would probably depend critically on requiring third party access across international borders in Europe. With a fully integrated pipeline grid with third party access a buyer in, say, Belgium could choose between Norwegian, Russian, Dutch, and Algerian deliveries, picking the one that would offer the best conditions. The Interconnector pipeline now under construction between the United Kingdom and the continent may come to play a very important role here, making it possible for the plethora of producers on the British continental shelf to offer gas to customers on the continent.

In this more realistic scenario of more than one supplier to the pipeline companies, it is likely that the gains from third party access would also be at the expense of the suppliers. If a pipeline company has access to supplies from more than one source, it would in the simplest of all conceivable cases buy only from the cheapest source, and the cost of gas from the next cheapest source would effectively put a limit to how much the cheapest supplier could get for his gas. In the real world things are not that simple. Supplies from one single source are not unlimited, and the buyers want to spread their risk by buying from several sources. Still the existence of alternative sources would appear to limit the price the suppliers can take and to augment the profits of the pipeline companies correspondingly, as long as the latter are in a monopoly position vis-à-vis the domestic distribution companies. In the next section we consider a simplified model with two suppliers.

A MODEL OF THIRD PARTY ACCESS
WITH TWO SUPPLIERS

Suppose there are two sellers of gas, a low-cost seller and a high-cost seller. Let the unit cost of these be denoted by C_l and C_h, respectively. There is, to begin with, a single pipeline company buying the gas and selling it on to the final consumers. For a numerical illustration we shall use the linear demand function

$$P = a - bQ. \tag{3.7}$$

For simplicity, we shall ignore the transportation costs and regard P as the price net of these costs.

Why would the pipeline company buy gas from two different suppliers? Would it not be most profitable to buy gas from the cheapest source, as the cheap supplier could be pressed to charge no more than the cost of the high-cost supplier, under the threat of losing the business to the high-cost supplier otherwise? So it might seem, but gas is a commodity that requires elaborate facilities and long pipelines, sometimes crossing several international borders. Technical problems and political turmoil are liable to cause supply interruptions, which would be inconvenient to say the least, the more so the more important gas is as a source of energy in the buying country. Security of supply has, therefore, been high on the agenda of gas buyers. While the Soviet Union still existed the countries of Western Europe were advised by the International Energy Agency not to rely on the Soviet Union for more than 30 percent of their supplies. Political considerations aside, countries may find it prudent not to rely wholly on one single source, as technical problems and even internal political troubles can afflict the most impeccably friendly nation.

Let us assume, therefore, that no more than a certain share α is desired from the low-cost supplier, despite his ability to offer better terms. And since he is the low-cost supplier, we may surmise that in fact the share α will be bought from this source. Hence we have

$$Q_l = \alpha Q \text{ and } Q_h = (1 - \alpha)Q, \; Q_l + Q_h = Q. \tag{3.8}$$

We assume that the share α is known to both sellers. In fact, the 30 percent maximum recommended in the days of the old Soviet Union was debated publicly and publicized. In a less polarized world, countries may be less candid about their policies and that, in fact, could be in their best interest when bargaining with their suppliers. How to deal with such less than perfect information is a challenging topic outside the scope of this book, so we proceed on the assumption that α is known.

The assumption that α is known and fixed makes it possible to proceed in a very simple manner akin to the approach in the previous section. The total profit of the gas purchase will be

$$P(Q)Q - \alpha QC_l - (1 - \alpha)QC_h, \tag{3.9}$$

as the transfer prices between the gas sellers and the pipeline company net out. Maximizing the total profit to be shared entails

$$P + P'Q = \alpha C_l + (1 - \alpha)C_h, \tag{3.10}$$

which is exactly the same kind of condition as (3.5) above, with the marginal cost of gas being replaced by the weighted marginal cost of the two suppliers, as the share of each is fixed by assumption.

The profit obtained by each supplier would be

$$\pi_l = Q_l(S_l - C_l), \; \pi_h = Q_h(S_h - C_h), \tag{3.11}$$

and the profit obtained by the pipeline company would be

Figure 3.4
Nash Bargaining Equilibrium with One Buyer (a Pipeline Company) and Two Sellers, a High-cost and a Low-cost Producer, with Fixed and Equal Supply Shares

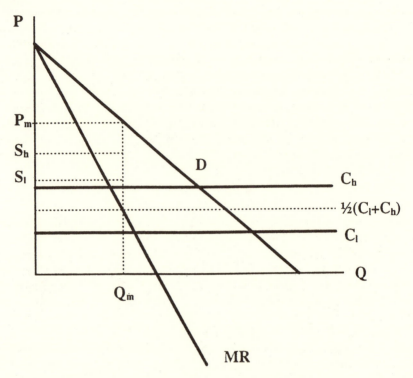

Note: C_l and C_h are the unit costs of production for the two suppliers, and Q_m is the profit-maximizing quantity, given the equal market shares. The producers receive the prices S_l and S_h, which divide the profit equally between the buyer and each seller.

$$\pi_T = Q_l(P(Q) - S_l) + Q_h(P(Q) - S_h). \tag{3.12}$$

Equal sharing of the profit due to each supplier between the pipeline company and the supplier entails.

$$S_l = C_l + \tfrac{1}{2}(P - C_l), \; S_h = C_h + \tfrac{1}{2}(P - C_h). \tag{3.13}$$

The solution is illustrated in Figure 3.4. The optimum monopoly price and quantity are P_m and Q_m, and the prices being paid to the two suppliers are S_h and S_l. Here the price being paid to the low-cost supplier happens to be higher than the cost of the high-cost supplier. This is possible because of the assumption of a fixed market share of each supplier. Relaxing this assumption and making the market share a part of the buyer's strategy would in all probability lower the prices the pipeline company pays; each supplier could be threatened with buying from the other one, which in the limit would cap the price at the cost level of the high-cost supplier. A formal consideration of this problem would make it necessary to consider, inter alia, the value of supply security (i.e., the disadvantage of extreme values of α).

Using the values $a = 10$, $b = 1$, $C_l = 2$, $C_h = 4$, and $\alpha = 0.5$ we arrive at the solution $Q = 3.5$, $P = 6.5$, $S_l = 4.25$, and $S_h = 5.25$.

What would happen under third party access? This depends critically on what strategy the two gas sellers are supposed to follow. A popular method of classifying strategies is by whether the price or the quantity sold is the decision variable. With price as a decision variable, the so called Bertrand competition, the low-cost seller would charge a price marginally below 4, and the quantity sold would almost double; it would become 6 instead of 3.5. If the demand curve truly reflects consumers' willingness to pay, this would go a long way toward attaining the optimum supply; if there is a practically unlimited supply at a cost of 4 per unit, the true marginal cost in the near term would be somewhere between 2 and 4, and there would not really be any place for the high-cost producer until the low-cost producer had exhausted his supplies, except for reasons of diversification.

The capacity to deliver gas is not likely, however, to be practically unlimited, and it is quite costly to build. It is therefore more likely that quantity will be the decision variable. With quantity as a decision variable each company would look at how much it should sell, for a given quantity sold by the other company. Here again there are two types of competition, the Cournot competition, where both parties make their decision simultaneously, and the Stackelberg competition, where one player has the advantage of being first to make the decision and takes into account how the other player will react. Here we shall look at the implications of the Cournot solution. The revenue (R) of supplier i will be given by multiplying the price (Equation 3.7) by the quantity sold by him (Q_i):

$$R_i = aQ_i - bQQ_i, i = l, h, Q = Q_l + Q_h. \tag{3.14}$$

The marginal revenue, given the quantity supplied by the other supplier, will be

$$a - bQ_j - 2bQ_i, i, j, = l,h; j \neq i. \tag{3.15}$$

The optimum solution for each supplier will be where his marginal revenue is equal to his marginal cost. A consistent solution is one where the quantity assumed by player i to be supplied by player j is in fact the quantity supplied by player j. Setting the marginal revenue (3.15) equal to the marginal cost and solving the equations gives as a consistent solution

$$Q_i = (a + C_j - 2C_i)/3b, i, j, = l,h; j \neq i. \tag{3.16}$$

For the above parameter values the solution is $P = 5.33$ and $Q = 4.67$, with $Q_l = 3.33$ and $Q_h = 1.33$. The high-cost producer would lower his supplies but the low-cost producer would expand his supply by more than that, so the total supply would increase and the price would fall. The price would still be higher, however, and the total quantity well below what would prevail under Bertrand competition. The low-cost producer's profit under Bertrand competition would be marginally below 12 while it would be less, or 11.09, under Cournot competition.

Would it not be natural, then, to expect Bertrand competition to prevail? Perhaps, but the exposition so far has ignored the important fact that the supply of gas requires the building of expensive production and transportation facilities. The simple models above are timeless, relating to flows per unit of time, whereas the supply of natural gas is limited, and bringing it to market requires large and irreversible investments. The following simple model is meant to throw some light on that issue.

Suppose our high- and low-cost suppliers have total gas deposits of 50 and 100 units, respectively. Their initial capacity is 1.75 each. Under the Cournot competition above the low-cost supplier would want to expand his capacity to 3.33. This he can only do at the cost of some additional investment in capacity. But capacity takes time to build. The present value of the profit from extraction for producer i can be expressed as

$$\int_0^{T_1} P(\textstyle\sum Q_{1,j})Q_{1,i}\, e^{-rt}\, dt + \int_{T_1}^{T_2} P(\textstyle\sum Q_{2,j})Q_{2,i}\, e^{-rt}\, dt$$

$$+ \int_{T_2}^{T_3} P(Q_{3,i})Q_{3,i}\, e^{-rt}\, dt - K_i \tag{3.17}$$

In this expression, future production is divided into three phases. During the first phase, up until T_1, the new capacity is being built. Production in this phase is Q_1. In the second phase the new, additional capacity is used and production is Q_2. At T_2 the producer with the lowest reserves/production ratio runs out of reserves, while T_3 is the time the reserve rich producer runs out. During this last phase the production is Q_3, determined by the capacity of the producer with the highest reserves/production ratio, so for the other producer this term is zero. K is the present value of the cost of the new capacity, and r is the instantaneous rate of discount.[8] Operating costs are ignored in this expression, but most of the cost of gas production is capital cost.

Now suppose each producer has the option to double his production capacity, from 1.75 prevailing when the pipeline company was the sole buyer and split its purchases evenly between the two suppliers, to 3.5, which is close to the optimal Cournot solution for the low-cost supplier. Suppose further that it takes five years to build the new capacity ($T_1 = 5$). Since the building of this additional capacity is a highly visible activity with a long lead time it is reasonable to suppose that both sellers can expand their capacity at the same time if they so wish. If only one of them wishes to expand his capacity the price will fall to 4.75, otherwise it will fall to 3. At the beginning of the final phase when only the reserve-rich producer is left producing the price will jump; it will be 6.5 if the low-cost producer is rich in reserves and chooses to expand his capacity, but otherwise it will be 8.25.[9] In the example to follow we shall assume that the low-cost producer has the highest reserve/production ratio, with initial reserves of 100, while the high-cost producer has only 50.

The cost of gas production is mainly the initial investment cost of production facilities and pipelines, etc. The figures 2 and 4 that we used in the static model above can be interpreted as annualized cost figures per unit, ignoring operating costs. At the prices 2 and 4, respectively, the producers break even. For $r = 0.1$ (approximately 10.5 percent rate of discount with annual compounding of interest) the present value at time zero of an annual production of 1.75 as long as the reserves last is 66 for the high-cost producer (the one with the low reserves) at a price of 4, and 35 for the low-cost producer at a price of 2. If production is just starting at time zero and the cost of doubling the capacity is the same as the cost of initial capacity, these would be the values of K for the two producers in expression (3.17). The present value of profits for the two producers (expression 3.17) and the two alternative strategies "invest" and "don't invest" is shown in the payoff matrix below, for $r = 0.1$.

Payoff Matrix

	No investment	**Investment**
No investment	107.2, 115.0	90.4, 103.9
Investment	48.6, 106.1	22.8, 86.1

Rows: Strategies of high-cost supplier; columns: strategies of low-cost supplier.

First figure: Payoff to high-cost supplier, second figure: payoff to low-cost supplier.

The dominant strategy in the payoff matrix is not to invest; either supplier always does best by not investing in any new capacity, irrespective of what the other does. Even if the pipeline company were obliged to transport the gas at a fair rate, the suppliers would not be interested in supplying any more, and the only effect of deregulation would be to deprive the pipeline company of its share in the rent.

This outcome depends critically on the cost of the new capacity. If the cost of the new capacity were lower, say, 20 and 10 respectively, we would get the payoff matrix below.

Payoff Matrix

	No investment	Investment
No investment	107.2, 115.0	90.4, 128.9
Investment	94.6, 106.1	68.8, 111.1

Rows: Strategies of high-cost supplier; columns: strategies of low-cost
  supplier.
First figure: Payoff to high-cost supplier; second figure: payoff to low-
  cost supplier.

In this case the low-cost supplier would choose to invest while the high-cost supplier would not. The best strategy for the high-cost supplier is not to invest, irrespective of what the low-cost supplier does. And irrespective of what the high-cost supplier does, the low-cost supplier does best by investing in new capacity. Hence the price will fall from 6.5 to 4.75 (and rise back to 6.5 when the high-cost supplier has run out of reserves), and the quantity supplied will increase to 4.25.

How does this compare to the optimum supply? The question of optimum supply is a tricky one with respect to nonrenewable natural resources and will be addressed more fully in Chapter 5. Briefly, an optimum supply schedule requires that the price of the nonrenewable resource, in this case gas, should rise over time and reach the level at which gas can no longer compete with substitutes at the time point when we run out of reserves. How quickly the price should rise depends on how large a reserve we have and how long a time it will take to deplete it. If reserves are sufficiently scarce, the optimum price of gas will be much higher than its marginal cost, reflecting its scarcity value. But reserves are being discovered continuously and no one knows how much there remains to be discovered. Calculations of optimal price paths therefore are purely hypothetical. The upper limit to the price of gas might be set by the cost of virtually unlimited supplies of gas, which is perhaps not much higher than the cost of contemporary high-cost producers.

The case for deregulation of the European gas market is probably stronger than this duopoly example might lead us to believe. The undeveloped gas reserves in Russia are huge, and a lot also remains to be developed in Algeria and the Middle East. The European gas market would not be a duopoly after eliminating the market power of the pipeline companies; there are at present four large players, Norway, Russia, Algeria, and the Netherlands, and there is a possibility that suppliers on the British continental shelf will enter the market when the Interconnector pipeline is finished, and possibly even suppliers in the Middle East. The more suppliers there are, the greater the competition will be, which means lower prices and larger quantities. Bjerkholt, Gjelsvik, and Olsen (1990a) have studied the European gas market, applying a model with ''lumpy'' capacity investments similar to the ones discussed above and three major suppliers (Norway, Algeria, and Russia). They predict a considerable increase in gas supplies and falling prices.

PRICES AND CONTRACTS IN THE EUROPEAN
GAS MARKET

Traditionally the natural gas market was characterized by bilateral negotiations and long-term contracts. Deregulation is changing this in the United States and the United Kingdom, with the emergence of spot markets and long-term contracts fading into the background or even being reneged on. Historically there were strong reasons for bilateral, long-term contracts; the high capital investment in gas extraction and transportation and the absence of spot markets made it necessary to secure financial flows to extraction and pipeline companies alike. The first contracts in the Norwegian part of the North Sea were depletion contracts that were valid for particular fields. Furthermore, contracts had take-or-pay clauses, specifying that the buyer would have to pay regardless of whether or not he wanted all the flow of gas he had contracted for.

Both depletion and take-or-pay contracts have fallen out of fashion in the aftermath of the deregulation of the markets in the United States and the United Kingdom. In the Norwegian part of the North Sea depletion contracts are no longer concluded. Later contracts for gas deliveries do not specify where the gas is to come from; as more fields have been developed gas can be provided from many different fields, and it is in fact advantageous to have some degree of freedom as to where the gas is to come from. If, say, gas associated with oil in what is predominantly an oil field has to be gotten rid of, it may be highly advantageous to use it to meet existing obligations for deliveries and slow down the depletion of a gas field instead; the alternative would be flaring the gas, which would be a waste and is in fact forbidden in the Norwegian part of the North Sea, or injecting the gas back into the field, which is expensive. Reinjection may, however, sometimes be desirable in its own right to enhance oil recovery, but that is a different question.

In the continental European market long-term bilateral contracts are still the rule. The terms of these contracts are a well-guarded secret, but it is nevertheless well known that the prices stipulated in these contracts usually are tied to the prices of oil products. After the two oil price hikes in the 1970s suppliers became interested in ensuring that the price of gas would follow the price of oil, which in those days was generally expected to move upward. Figure 3.5 shows the development of gas prices exported through pipelines from the Netherlands and Norway, together with the price of Brent Blend, for the period 1980–94. The price of gas exported from the Netherlands increased gradually from 1980 to 1982, probably due to contracts that tied the price of gas to the preceding development in the price of oil (as discussed in Chapter 1, the price of oil rose steeply in 1979–80). The steep fall in oil prices in the first half of 1986 dragged

Figure 3.5
Prices of Natural Gas and Crude Oil, 1980–94

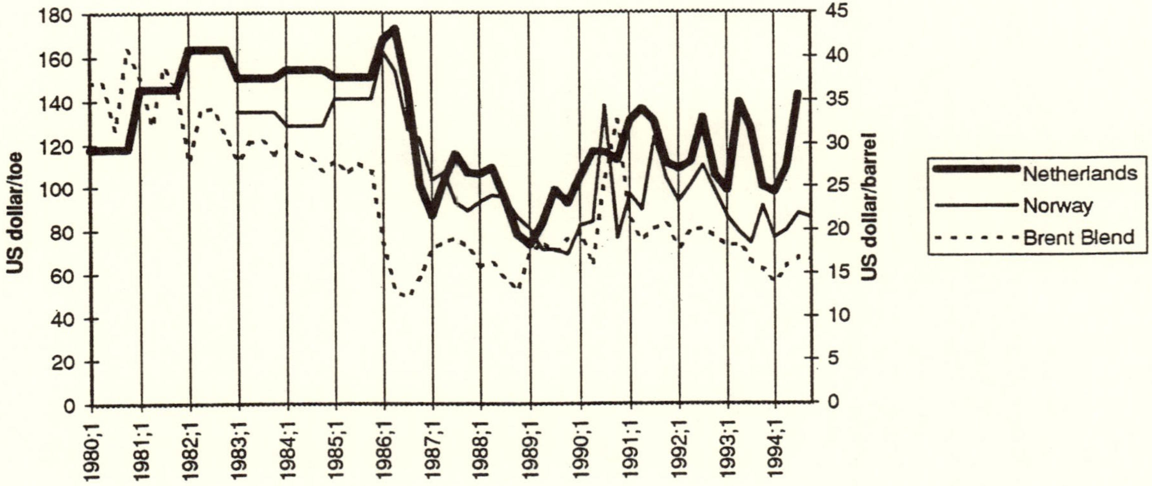

Legend: Netherlands, Norway, Brent Blend

US dollar/toe — 180, 160, 140, 120, 100, 80, 60, 40, 20, 0

US dollar/barrel — 45, 40, 35, 30, 25, 20, 15, 10, 5, 0

X-axis: 1980:1, 1981:1, 1982:1, 1983:1, 1984:1, 1985:1, 1986:1, 1987:1, 1988:1, 1989:1, 1990:1, 1991:1, 1992:1, 1993:1, 1994:1

Source: Statistics Norway, Oil and Gas Activity.

down gas prices with a lag; the prices of Norwegian and Dutch gas fell equally dramatically from the first half of 1986 to the first quarter of 1987. The recovery of oil prices in 1987 appears to have pulled the Dutch price upward again, and possibly the Norwegian price as well but with a longer time lag. The new dip of oil prices from 1987 to late 1988 dragged the price of both Dutch and Norwegian gas down again. Since then gas prices have been quite volatile but they have broadly followed the price of oil, particularly the Norwegian price. Whatever the reason, the prices of Norwegian and Dutch gas can apparently deviate substantially for some period. As for oil, the price level of gas has been decidedly lower after the drama of 1986, a development that would be much more marked if we had deflated the price series.

But why tie the price of gas to oil? Gas would most likely conquer market shares from oil if it were not tied to a rising price of oil. Furthermore, gas is not a substitute for oil in all uses. Gas is a substitute for coal, besides oil, in the generation of electricity, and indirectly for both of these when it substitutes for electricity in applications such as home heating or cooking. Apparently the gas-selling nations have been more preoccupied with not selling their gas cheaply than with conquering new markets. The multitude of possible uses of gas have, however, led to corresponding variations in the indexation of contracted gas prices; the price is sometimes indexed to heavy fuel oil, which is a substitute in industry and power generation, to middle distillates such as gas oil, which is a substitute in home heating, or to coal, which is a substitute in electricity generation, or to the price of electricity itself, and in fact usually to a combination of these. These combinations vary according to the end use for which the gas is chiefly intended. Furthermore the price of gas is sometimes indexed to the rate of inflation.

NOTES

1. On the alternative *abyssal, abiotic* theory, see footnote 2 to Chapter 1.

2. The peak demand (on a monthly basis) for natural gas in OECD Europe is more than twice the lowest demand (OECD, 1994, p. 35).

3. The cost is, presumably, proportionate to the circumference, and the relationship between circumference (c) and diameter (d) is $c = \pi d$. One would expect the capacity to be proportionate to the area (A) contained within the circumference, which is related to the diameter by $A = \pi d^2/4$. However, according to OECD (1994, p. 46), the capacity of a pipeline is proportionate to its diameter raised to the power of 2.5. This is due to less friction with the walls in large pipes.

4. Other phrases used are open access and common carriage. Some authors make a distinction between these two, using common carriage for the situation where a pipeline is obliged to service any customer (but must ration capacity in a fair way if it is insufficient).

5. We ignore the fact that pipeline companies in fact sell the gas to local distributors who themselves are in a monopoly position vis-à-vis the final consumers, except for large, industrial buyers. The transfer price of gas from the pipeline companies to the

local distributors is in fact subject to negotiations between the two and so could be considered in a separate bargaining model (cf. Vislie, 1990). That notwithstanding, it is highly likely that the pipeline company will face a downward-sloping demand curve, even if the price does not correctly reflect consumers' marginal willingness to pay.

6. The first derivative is denoted by a prime.

7. This is the so-called Nash bargaining solution (Nash, 1950).

8. In continuous time, the discount factor is e^{-rt} instead of the more familiar $(1 + r)^{-t}$ applied in discrete time. If compounded n times per year, one monetary unit would grow to $(1 + r/n)^{nt}$ over t years. We can write this as $(1 + 1/(n/r))^{(n/r)rt}$. As $n \rightarrow \infty$ and interest is compounded continuously, the number $(1 + 1/(n/r))^{(n/r)}$ converges to e, the basis of the natural logarithms. Hence the continuous time analog of compound interest, $(1 + r)^t$, is e^{rt}. Needless to say, for the same r, $(1 + r)^t < e^{rt}$. This explains perhaps why banks typically add interest to their depositors' accounts once a year but charge interest on their loans every month or quarter.

9. Would such a jump in price be compatible with a market equilibrium? Not in a world with full certainty and reversible decisions; the reserve-poor, high-cost producer would foresee the future jump in price and withhold some production for a later day. But the high-cost supplier has already built his capacity, and he will have interest and amortization payments to attend to, and his shareholders are likely to demand dividends. The production capacity has to be maintained, and such maintenance costs are likely to make it unattractive to withhold production.

Chapter 4

Some Principles of Petroleum Production

THE PHYSICAL LAWS OF PETROLEUM PRODUCTION

Oil and natural gas are stored in underground reservoirs. These reservoirs are porous rock formations where the petroleum is contained in the pores and prohibited from seeping upward to the ground by a solid rock putting a lid, as it were, on the reservoir. Prospecting for oil by geological surveys involves identifying possible structures containing and trapping petroleum.

Since these petroleum-containing structures are several kilometers below the surface, the pressure in the reservoirs is much higher than atmospheric pressure. Drilling a hole into such reservoirs is therefore like punching a hole in a balloon; as air will fizzle out of the balloon, so oil or gas will guzzle out of the reservoir. Drilling for oil therefore involves precautionary measures (valves and fluids in the borehole) that prevent the oil from streaming out of the reservoir in an uncontrolled fashion. Sometimes this is unsuccessful, which gives rise to so-called blowouts involving spectacular geysers of black liquid, which in the worst of cases catch fire.

Utilizing this natural flow of the oil out of the reservoir is how oil is produced, in the simplest of cases. It is customary to distinguish between three main sorts of driving mechanisms, or just "drives," which bring the oil out. Basically, what is involved is that as the pressure falls in the reservoir and the oil flows out, the emptied pores in the rock must be filled with something else. It is that "something else" that distinguishes the different "drives."

Dissolved Gas Drive

Petroleum is a mixture of heavy and light products, the heaviest consisting of molecules with many carbon atoms. The lightest products are the ones that make up natural gas. These are gaseous under normal temperatures and atmospheric pressure but may be liquid under the high pressure prevailing in the reservoir. Often the natural gas is mixed with oil in the reservoir. As the pressure falls in the reservoir the natural gas expands, and liquid natural gas becomes gaseous and expands, driving the oil toward the wells where the pressure is lowest.

Gas Cap Drive

Sometimes natural gas, being light, is concentrated in the uppermost part of the reservoir, underneath the solid rock that prevents the oil from seeping through the ground. As crude oil is extracted from the lower part of the reservoir, the gas cap expands, driving the oil toward the wells.

Water Drive

Underneath the oil in the reservoir there is groundwater. As oil is extracted, the groundwater seeps into the reservoir and fills the pores that contained the oil. To maximize the recovery of oil it is important to control the rate at which the pressure falls, for otherwise water can bypass some of the oil, leaving it trapped and inaccessible in small pockets here and there.

A typical time profile of petroleum production from a reservoir is shown in Figure 4.1. Initially the production increases; wells being drilled and successively put into production. After the desired number of wells has been drilled, the production reaches a plateau where it typically stays until about one-half of recoverable reserves have been tapped from the reservoir.[1] Thereafter the rate of production declines and is discontinued when the value of current production no longer covers the cost of operating the production equipment. We can recognize these three phases in the development of the production rates from some Norwegian fields shown in Figure 4.2. One shows the Ekofisk area, which contains several reservoirs. There is a single peak but no plateau, and instead of continuing to decline the production rate rose after 1987, due to a water injection project for enhancing the production. The Murchison field has no plateau either but a single and distinct peak, from which it declined rapidly. The Statfjord area had a long development phase and also a long plateau phase and has recently started to decline. The production profile from the Frigg area has the shape of a hump. Frigg is mainly a gas field and the production profile is influenced by contracts for selling the gas.

There are two reasons why the rate of production goes into decline after a while. First, as oil recovery progresses, the pressure falls in the reservoir. This ultimately slows down the rate at which the oil is driven toward the wells. This is particularly true of reservoirs with dissolved gas drives. Second, as the water table creeps into a reservoir with a water drive the water blends with the oil, and the well starts producing a mixture of oil and water. Some wells that have

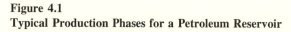

Figure 4.1
Typical Production Phases for a Petroleum Reservoir

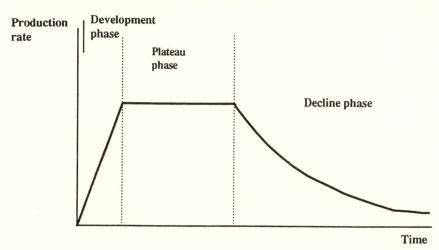

Note: Production increases gradually while additional holes are being drilled and the production capacity is built up. The production rate then remains stable, with the production equipment being used to full capacity. Finally production tapers off when the pressure in the reservoir falls, or as groundwater seeps in and becomes mixed with the oil.

gone into decline in the British sector of the North Sea produce about as much water as oil.

As a reservoir goes into a decline it may be worthwhile to inject water or gas into it to increase the pressure. This is usually called secondary recovery. In the North Sea and elsewhere water and gas injection are often done right from the beginning, to ensure an optimum utilization of expensive extraction installations. Other methods for enhancing production, such as injection of chemicals, may supersede water and gas injection and are referred to as tertiary recovery. It may also be necessary to pump the oil out of the ground if the pressure in the reservoir has become too low to lift it up.

THE DECLINE IN THE PRODUCTION RATE

The rate at which the production of oil falls is determined by physical factors such as the viscosity of the oil, the permeability of the rock in which it is stored, the rate at which the pressure in the reservoir falls, how fast water seeps into the reservoir, and so forth. It is possible to model all these things mathematically. This is the subject of reservoir engineering, an applied science used by the oil companies to investigate various possible development patterns of the production from a reservoir and help decide how many wells to drill, where to drill, and so forth. Needless to say, the physical factors are not known with certainty.

Figure 4.2
Production Profiles from Four Norwegian North Sea Oil and Gas Fields

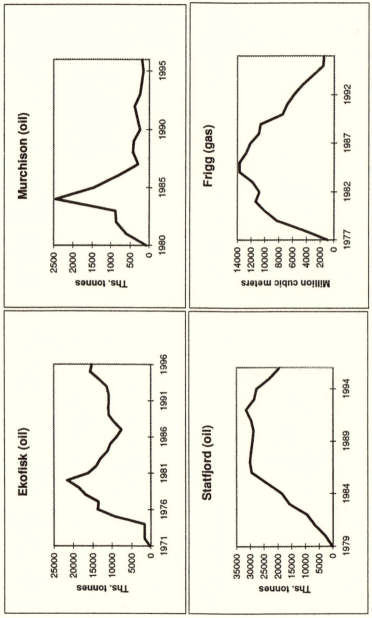

Note: The profiles show either oil or gas, as indicated, but all produce some gas in addition to oil, or vice versa. Ekofisk and Frigg comprise areas with several fields. The increase in production on Ekofisk is due to enhanced recovery from water injection, which started in the mid-1980s. Figures for Murchison and Statfjord show the Norwegian share.

Source: Statistics Norway, *Oil and Gas Activity.*

For economic purposes some mathematically simple decline curves are used to predict the possible development of production from a reservoir. The most popular ones are of the family of "general hyperbolic decline," the mathematical expression of which is[2]

$$(dq/dt)/q = -1/(a + bt),$$ (4.1)

where a and b are constants. If $b = 1$ we get the so-called harmonic decline curve while $b = 0$ gives rise to exponential decline where production declines at a constant rate $k = 1/a$. Decline curves with different values of b are shown in Figure 4.3.

Note that a continuous exponential decline would result if the flow rate of oil is proportionate to the pressure in the reservoir and the rate at which the pressure falls is proportionate to the flow of oil. In that case we would have

$$q_t = k_1 h_t,$$ (4.2a)

$$dq_t/dt = k_1 dh_t/dt,$$ (4.2b)

$$dh_t/dt = -k_2 q_t,$$ (4.2c)

where q_t is the flow rate of oil at time t, h_t is pressure at time t, and k_1 and k_2 are constants. This gives

$$dq_t/dt = -kq_t,$$ (4.3)

where the rate of production declines at the constant rate $k = k_1 k_2$. This differential equation has the solution

$$q_t = q_0 e^{-kt},$$ (4.4)

where q_0 is the initial rate of production. Note that k, the rate of decline, depends on the initial rate of production (q_0), even if it is constant over time once the initial rate of production has been determined. From the above we have

$$k_1 = q_0/h_0.$$ (4.5)

The initial pressure in the reservoir, h_0, is given by nature, while q_0, the initial rate of production, depends on how many wells are drilled, the rate at which oil is pumped out of the wells, and so forth. Hence the decline rate, k, is

$$k = k_2 q_0/h_0,$$ (4.6)

and we are faced with a tradeoff between the initial rate of production and the rate at which it declines. This has important economic implications, given that

Figure 4.3
Hyperbolic Decline Curves

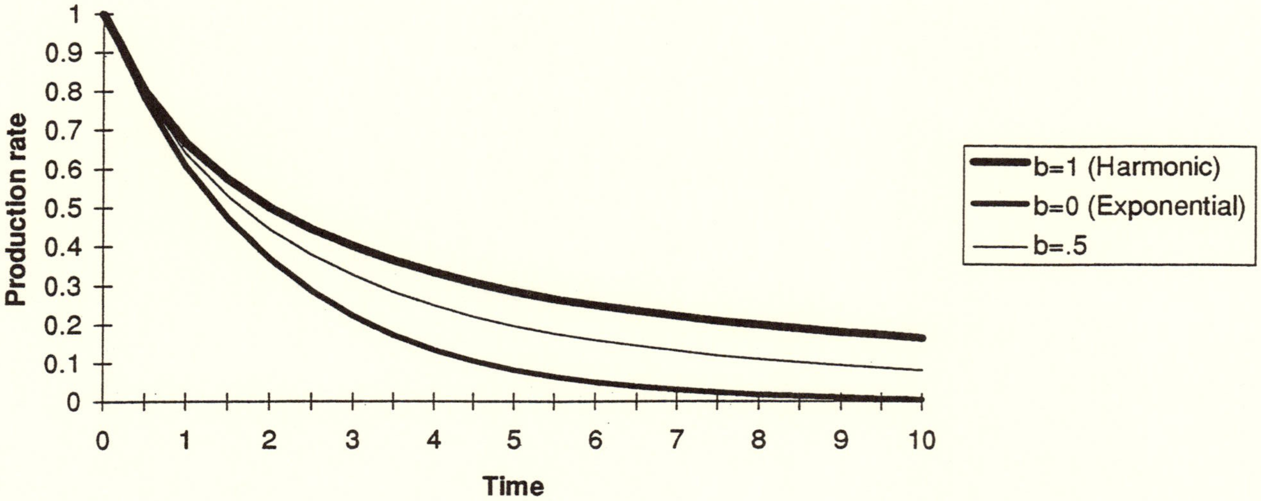

Note: The rate of production (q) falls according to the equation $(dq/dt)/q = -1/(a + bt)$.

future incomes are discounted vis-à-vis present ones, to which we shall return below.

For economic purposes, the exponential decline will often be a good enough approximation, and it has the advantage of simplicity. The time profile of decline is affected by the geological characteristics of the reservoir and the fluidity of the oil (or gas), but empirical decline profiles are rarely smooth. The exponential curve will nevertheless often be a good approximation, particularly when calculating present values of future income streams; deviations from the curve will cancel each other out.

RATE SENSITIVITY

In addition to affecting the rate of production decline, the initial rate of production (q_0) may also affect the total amount of oil that can be recovered from a reservoir. Two effects are, or can be, involved. First, a higher initial production rate and the resulting drop in pressure may cause water flooding of the reservoir; that is, the water would seep quickly into the reservoir and bypass or "trap" some of the oil, preventing it from ever getting to the wells. This would reduce the total amount of recoverable reserves. Second, more production wells can result in a more efficient draining of the reservoir. This would augment recoverable reserves and thus offset the rise in the decline rate that results from increasing the initial production rate.

The reason for the latter effect is that oil and gas do not flow without friction in their underground reservoirs. The rate at which they migrate toward the wells depends on the permeability of the rock in which they are trapped and the viscosity of the oil itself. The thicker the oil and the more dense the rock, the lesser the rate of migration. Few, if any, reservoirs could be tapped satisfactorily with a single well; oil from the more distant parts of the reservoir would not flow toward the well. Therefore, the recoverable reserves will increase with the number of wells drilled but, as will be demonstrated below, at a decreasing rate. Ultimately the disadvantages of too rapid pressure fall will take over and cause recoverable reserves to decline.

How recoverable reserves may increase with the number of wells drilled (and the initial production rate, assuming that all wells are drilled simultaneously at the beginning, which is a simplification) may be demonstrated as follows. Suppose the portion of oil in place at each location that can be recovered by each well depends on the distance from the well. The amount of oil recovered at each location (y) can then be written as a function of the distance (x) from the well:

$$y = f(x). \tag{4.7}$$

Decreasing drainage with distance implies $f'(x) < 0$. The area drained by each well will be approximately A/N, where A is the size of the entire field and N is the number of wells. The distance *(r)* to the edge of the draining area of each well, which we approximate by a circle, will be

$$r = \sqrt{\frac{A}{\pi N}} \tag{4.8}$$

where π has the usual geometric meaning. The production from each well *(q)* will be

$$q = \int_0^r 2\pi x f(x)dx = \pi r^2 f(r) - \int_0^r \pi x^2 f'(x)dx \tag{4.9}$$

To see how the total production from the area changes with the number of wells we calculate the derivative

$$\frac{d(Nq)}{dN} = q + N\frac{dq}{dN} = -\int_0^r \pi x^2 f'(x)dx \tag{4.10}$$

Total production will increase with the number of wells drilled only if $f'(x)$ < 0, that is, if the recovery rate increases as we get closer to the well. If the recovery rate is constant it would mean that the entire reservoir could be depleted by just one well. Each additional well would only steal oil from other wells, as it were, and would only change the time profile of production, as in the examples to be discussed below. Sooner or later, however, the marginal increase in production by drilling more wells will decline, because the value of the integral term declines as r declines, for any given value of $f'(x)$, and as r approaches zero and N approaches infinity the integral will vanish.

ECONOMIC ANALYSIS

We shall focus here on two basic economic choices in petroleum production, how much to invest in production capacity and when to terminate production. Production capacity is determined by the number of wells drilled, equipment for treatment, and so forth. This determines the plateau production rate and the rate of decline; the higher the plateau the shorter it is, and the more quickly the production rate will decline. This is essentially a choice of a time-optimal path of production. We shall use the exponential decline function to illustrate the basic points, which implies that we ignore the plateau phase. This is a simplification and does not affect the qualitative conclusions regarding, for example, the effect of the interest rate on the optimum investment level or whether or not to go for maximizing the recoverable reserves. In addition, the initial rate of production may affect the amount of recoverable reserves, as already explained. This will also be considered below.

Case I: Constant Production Rate

The simplest possible model for depleting an oil reservoir is one where the rate of production is constant over time, so the production stays on the plateau and never declines. This is patently unrealistic but useful to bring out a fundamental result, namely that the interest rate has an ambiguous effect on the level of investment in production capacity. Let c be the cost per unit of production capacity, measured as the initial (and here constant) production rate, q_0. Let r denote the instantaneous rate of discount.[3] With investment occurring at time point 0 and production starting immediately, the net present value (V) of production will be

$$V = -cq_0 + \int_0^T pq_0 e^{-rt} dt = -cq_0 + pq_0(1 - e^{-rT})/r, \tag{4.11}$$

where p is the (constant) price of the oil produced and T is the time at which the production comes to an end.

In this case with a constant production rate, all the oil will be extracted from the reservoir, provided it is economical to extract anything at all. The time at which the production will be ended is determined by the requirement that the accumulated extraction be equal to the total available reserves when production began. Since the production rate is constant, this is simply

$$T = Q/q_0, \tag{4.12}$$

where Q is the reserve available initially. Note that changing the initial production rate only affects the time it takes to deplete the reservoir; the entire reserves could be recovered through just one well, as it were. This is unrealistic, but the assumptions of no decline in production and no rate sensitivity will be relaxed below one at a time and the consequences examined.

We can now find the optimum level of investment (q_0) by setting $dV/dq_0 = 0$.[4] This gives

$$(1 - e^{-rT})/r - Qe^{-rT}/q_0 = c/p. \tag{4.13}$$

To see the effect of a higher rate of interest on the optimum level of investment, we can differentiate the left-hand side of this expression with respect to r and q_0 and set the result equal to zero, since the right-hand side is a constant. Making these calculations shows that the differential quotient dq_0/dr has an ambiguous sign, implying that an increase in the rate of interest can either lower or raise the optimal level of investment. The reason why this happens is that an increase in the rate of interest has two opposite effects. On the one hand, it decreases the present value of future production, making it less attractive to invest in developing the field. On the other hand, a higher rate of interest means that incomes in the near future are more valuable than incomes in the more distant future, but to produce more in the near term it would be necessary to increase the investment and the initial rate of production.

Another way of expressing this is by saying that the rate of interest plays a dual role in the economy. On the one hand, it expresses time preference, so a higher rate of interest means that production should be accelerated. On the other hand, it is an expression of the opportunity cost of capital, so a higher rate of interest makes it more costly to invest in production equipment as necessary to accelerate the production. Not surprisingly, the latter effect dominates when the cost of investment in production capacity is relatively high (a high c/p ratio). The ambiguous effect of the interest rate is illustrated in Figure 4.4.

Case II: Declining Production Rate, No Rate Sensitivity

We move now to the more realistic case where production declines over time. We shall use the exponential decline function, as this provides the simplest case for illustrating the principles to be discussed. Since the production rate is now falling over time, it is important to consider explicitly the operating costs, since production will be terminated when it is no longer worthwhile to keep it going. Operating costs are likely to be approximately proportionate to capacity (q_0). Denoting the unit operating costs by a and making use of (4.4) above, the net present value of production from the field is

$$V = -cq_0 + \int_0^T q_0(pe^{-kt} - a)e^{-rt}dt \qquad (4.14)$$
$$= -cq_0 + pq_0(1 - e^{-(k+r)T})/(k + r) - aq_0(1 - e^{-rT})/r,$$

where k is the decline rate of production.

A straightforward hypothesis is that production will be terminated when it no longer makes any positive contribution to the income stream (circumstances that might modify this conclusion will be discussed below). This implies $pe^{-kt} = a$, or

$$T = -\ln(a/p)/k.$$

Note that here the time horizon of production depends on the initial capacity because k, the decline rate, depends on initial capacity (see [4.6] above):

$$T = -\ln(a/p)h_0/k_2q_0, \qquad (4.15)$$

Figure 4.4
Effect of the Discount Rate (*r*) on the Initial Rate of Production (q_0) for Two Levels of Unit Cost of Initial Capacity (*c*)

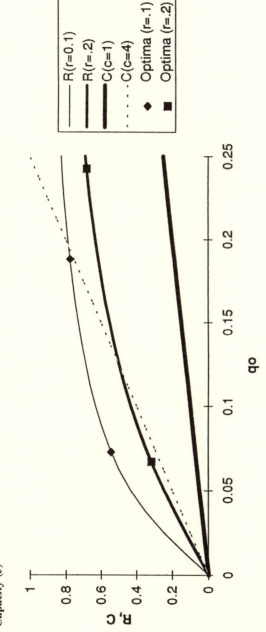

Note: R = present value of revenues; C = investment cost. The two optima shown for each value of *r* correspond to a high and low unit cost of investment (*c*), respectively.

so the production phase will become shorter as the initial production capacity increases.

Substituting T into (4.14) and taking the derivative dV/dq_0 and setting it equal to zero makes it possible to find the optimum initial capacity and how it changes as a result of a change in the discount rate. A change in the discount rate has an ambiguous effect on the optimum initial rate of production, as in the case with a constant rate of production. This is illustrated in Figure 4.5. For a "low" unit cost of investment ($c = 1$), raising the discount rate from 0.1 to 0.2 increases the optimum initial rate of production (q_0) from 2.17 to 2.54, while for a "high" unit cost of investment ($c = 2$) the same change in the discount rate lowers the optimum initial rate of production from 1.27 to 1.17.

A curious effect in models with an exponential decline is that the total amount produced is determined by operating costs and price but is independent of the discount rate and the cost of initial capacity. The reason is that the entire reserves could be recovered through just one well, as in the simple oil tank model of Case I above. The rate of interest and the investment cost (c) affect the initial rate of production (q_0) or how many wells to drill. This again affects the time profile of production. Operating costs affect the total amount produced because they affect the length of time the wells should be kept operative. To demonstrate this, note that the total production is simply the sum of production levels in each period or, in a continuous time model, the integral of the production rate over the period of production. Using (4.4), (4.6), and (4.15) gives

$$\int_0^T q_t dt = \int_0^T q_0 e^{-kt} dt = \frac{q_0(1 - e^{-kT})}{k} = \frac{h_0(1 - e^{\ln(a/p)})}{k_2} = \frac{h_0(1 - a/p)}{k_2} \qquad (4.16)$$

Using, for example, the parameter values $k_2/h_0 = 0.05$, $p = 1$, and $a = 0.2$ gives a total extraction of 16.

Before going on to consider rate sensitivity it is convenient to further discuss the optimum shutdown time of a field in the context of the exponential decline model with given recoverable reserves that are independent of the initial production rate and how it declines over time. There are two economic considerations in addition to operating costs that in particular impinge on the optimum shutdown time: the dismantling costs and the option value of keeping the field accessible in case its profitability improves in the future, through either rising oil prices or falling costs.

Dismantling Costs

Removing production installations for offshore oil and gas fields is not a trivial undertaking, as exemplified by the Brent Spar episode[5] in the summer of 1995. Offshore platforms in places such as the North Sea are huge installations and expensive to construct, and also expensive to dismantle. The impact of

Figure 4.5
The Effect of the Discount Rate on the Present Value (V) of a Field and the Optimum Initial Rate of Production (q_0)

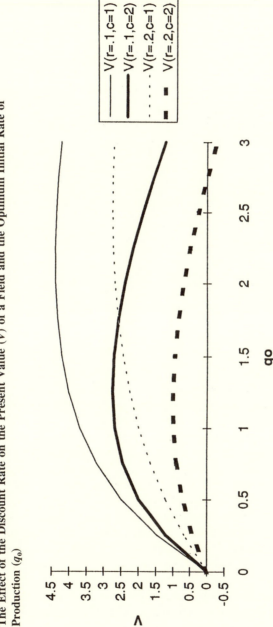

Note: The optimum initial rate of production is given by the top of the present value curve. For a "low" unit cost of investment ($c = 1$), raising the discount rate from 0.1 to 0.2 increases the optimum initial rate of production from 2.17 to 2.54, while for a "high" unit cost of investment ($c = 2$) the same change in the discount rate lowers the optimum initial rate of production from 1.27 to 1.17.

Table 4.1
Numerical Example of the Impact of Dismantling Costs (b) on the Optimum
Shutdown Time (T) (parameter values: $r = 0.1$, $p = 1$, and $c = 2$)

b	q_0	T	V
0	1.27	25.35	2.766
0.5	1.2	38.38	2.723

dismantling costs can be shown formally as follows. Suppose dismantling costs, like construction costs, are proportionate to the initial production capacity, q_0. Denote unit dismantling costs by b. The net present value of production will then be given by a modified version of (4.14):

$$V = -cq_0 + \int_0^T q_0(pe^{-kt} - a)e^{-rt}dt - bq_0e^{-rT} \tag{4.14'}$$
$$= -q_0 (c + be^{-rT}) + pq_0(1 - e^{-(k+r)T})/(k + r) - aq_0(1 - e^{-rT})/r.$$

To find the optimum shutdown time, take the derivative of V with respect to T and set it equal to zero:

$$dV/dT = q_0[pe^{-kT} - a + rb]e^{-rT} = 0$$

From this we can find T, the optimum shutdown time, as

$$T = \{ln[p/(a - rb)]\}/k. \tag{4.15'}$$

The effect of dismantling costs is to prolong the period of production. Postponing dismantling costs gives a capital gain of $rbq_0\Delta t$ over a "short" time period Δt (imagine that the producer has set aside a sum bq_0 to cover dismantling costs and can earn an interest on this sum as long as it is deposited in a bank account or invested in the capital market). This justifies producing at an operating loss, as long as that loss, $q_0(pe^{-kT} - a)$, is less than the gain of postponing dismantling costs, rbq_0. An illustration of this is provided by the numerical example in Table 4.1.

In the example in Table 4.1 the dismantling costs are almost negligible in terms of the present value of production but nevertheless have a considerable effect on the optimum shutdown time. The latter is in fact a combination of two effects. First, as already explained, the postponement of dismantling costs justifies producing at an operating loss for some time. Second, dismantling costs make it less attractive to invest in production equipment, which lowers the initial production rate q_0. But lowering the initial production rate reduces the decline rate and extends the period of production, as already explained.

Dismantling costs also have the somewhat curious effect of increasing the total amount extracted, owing to their prolongation of the extraction. Total extraction is given by

$$\int_0^T q_t dt = \frac{q_0(1 - e^{-kT})}{k} = \frac{h_0(1 - e^{\ln[(a - rb)/p]})}{k_2} = \frac{h_0(1 - (a - rb)/p)}{k_2} \tag{4.16'}$$

Using the same parameter values as in Table 4.1 and the numerical example above, we get 17 instead of 16.

The Option Value of Continuing Production

Installations for offshore oil and gas production are elaborate; the largest production platforms in the North Sea are taller than the Eiffel Tower. Once production has been discontinued and maintenance work abandoned it would be expensive to start producing again, and even more so if the structures have been removed. Production shutdown may therefore be regarded as irreversible. As we have seen, production will be shut down when the current income from production plus the rate of return on postponing dismantling costs have fallen to a level equal to the operating cost. But the current revenue depends on prices. What if prices can be expected to rise in the not too distant future? Production might then become profitable again and the producer would regret having foreclosed the option of continuing production.

If there is a sufficiently high probability that the price will rise in the not too distant future there is a net expected gain to be realized from continuing to produce at a loss for some time in the hope of taking advantage of a higher price later. The net present value of expected profit from continuing production represents the value of the option of maintaining production. It is noteworthy that this value may be positive even if the expected price is equal to the current price at which it would be advisable to shut down production if the latter were expected to remain constant; it turns out that the upside gain from a higher price is greater than the downside loss from a lower price. The same reasoning would apply to uncertainty with regard to operating costs.

Let us illustrate with a numerical example, using the exponential decline model. We ignore dismantling costs and fix the following parameters: $q_0 = 1$, $k = 0.05$, $p = 1$, $a = 0.2$, and $r = 0.1$. The optimum shutdown time would be at $T = 32.1887$, ignoring the option value of continuing production. The price is, however, uncertain. Suppose it is expected to rise to $p_1 > p$ with probability s, and to fall to $p_2 < p$ with probability $1 - s$ at time $T + \Delta T$. The option value (O) of continuing production beyond T will be

$$O = V_1 + sV_2(p_1) + (1 - s)V_2(p_2), \tag{4.17}$$

where V_1 is the net present value of continuing production over the time period ΔT at the end of which the price is expected to change, and V_2 is the net present value of production after the price has changed, assuming that an optimum shutdown time will be selected at time point $T + \Delta T$. V_2 can assume two values, depending on the price obtaining after $T + \Delta T$. If the price turns out to fall instead of rising there will be no point in continuing production, so $V_2(p_2)$ will be zero. The expressions for V_1 and $V_2(p_1)$ are as follows:

$$V_1 = \int_0^{\Delta T} q_0(pe^{-k(T+t)} - a)e^{-rt}dt$$

$$= q_0 \left[\frac{pe^{-kT}(1 - e^{-(r+k)\Delta T})}{r + k} - \frac{a(1 - e^{-r\Delta T})}{r} \right] \qquad (4.18)$$

$$V_2 = \int_{\Delta T}^{T*-T} q_0(p_1 e^{-k(T+t)} - a)e^{-rt}dt$$

$$= q_0 \left[\frac{p_1 e^{-k(T+\Delta T)}(1 - e^{-(r+k)(T*-T-\Delta T)})}{r + k} - \frac{a(1 - e^{-r(T*-T-\Delta T)})}{r} \right] e^{-r\Delta T}$$

As an example, set $s = 0.5$, $p_1 = 1.5$, and $\Delta T = 2$. This gives $O = -0.0170 + 0.0702 = 0.0532$, a positive option value. Obviously, however, it will not pay to keep the production going if the price change is supposed to take place too far into the future. In this example, the option value will become negative for $\Delta T > 3.25$.

Case III: Declining Production Rate with Rate Sensitivity

As explained above, the total recoverable reserves (Q) are likely to depend on the initial rate of production (q_0), which again depends on the number of wells drilled. With few and widely spread wells the recoverable reserves are likely to increase as additional wells are being drilled, but with many and dense wells the pressure in the reservoir would be likely to fall too rapidly, leaving some oil trapped in inaccessible pockets, a problem that would be aggravated by an additional well. The relationship between the initial rate of production and recoverable reserves would then be as indicated by the Q curve in Figure 4.6. The upper limit to recoverable reserves (top of the Q curve) is most appropriately regarded as being determined by technological and geological constraints rather than the total amount of hydrocarbons contained in the reservoir.

For any given initial rate of production and recoverable reserves, the production rate will decline after a while. Let us again, for illustrative purposes, use the exponential decline model from the previous section. Again ignoring the plateau phase and assuming that the decline begins as soon as production has started, we can approximate the decline rate as follows. If production would go on indefinitely, the total amount recovered would be equal to the recoverable reserves, Q. That is

Figure 4.6
Recoverable Reserves (Q) and the Initial Rate of Production (q_0)

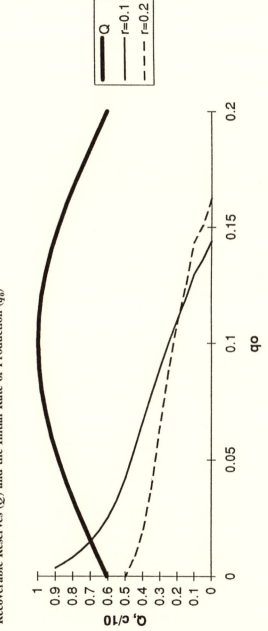

Note: The thick line illustrates how recoverable reserves from an oil field depend on the initial rate of production (related to the number of wells drilled), through the equation $Q = Q^* \exp(-\dfrac{(q_0 - q^*)^2}{2s^2})$, where $q^* = 0.1$ and $s = 0.1$ The thin lines show how the optimum initial rate of production depends on the unit cost of investment (c), for two different values of the discount rate.

$$Q = \int_0^\infty q_0 e^{-kt} dt = \frac{q_0}{k} \qquad (4.19)$$

from which we can find the decline rate as $k = q_0/Q$.

We can find the optimum initial rate of production in a way similar to above. The insight added by this particular model is that it does not necessarily make sense to maximize the recoverable reserves (get to the top of the Q curve in Figure 4.6); to do so requires investment in production equipment and the drilling of wells, so it might not be worthwhile. On the other hand, if investment costs are low, it might be worthwhile to overinvest in production equipment and reduce the recoverable reserves through overinvestment, in order to get the incomes early.

To see this clearly, let us look at the implications of maximizing the discounted value of production from the reservoir. To focus on the initial production rate and its dependence on investment costs and the rate of interest we shall ignore operating costs and dismantling costs. Production will then be maintained indefinitely, and the present value of profits will be

$$V = -cq_0 + \int_0^\infty pq_0 e^{-(k+r)t} dt = q_0[-c + \frac{p}{k+r}] \qquad (4.14'')$$

Maximizing with respect to q_0 gives

$$\frac{dV}{dq_0} = -c + \frac{p}{k+r} - \frac{p}{(k+r)^2} \frac{dk}{dq_0} \qquad (4.20)$$

where, from (4.19),

$$\frac{dk}{dq_0} = \frac{1}{Q} - \frac{q_0}{Q^2} \frac{dQ}{dq_0} \qquad (4.21)$$

As an illustration, let Q depend on q_0 as follows:

$$Q = Q^* \exp\left(-\frac{(q_0 - q^*)^2}{2s^2}\right) \qquad (4.22)$$

where q^* is the initial rate of extraction that would result in maximum recoverable reserves (Q^*) and s is a parameter governing how steeply the curve Q rises to its maximum (Q^*).

Figure 4.6 illustrates the kind of solutions we get for this problem. The thick

Table 4.2
Recoverable Reserves (Million Sm³ [Oil] and Billion Sm³ [Gas])

	1983		1996	
	Oil	Gas	Oil	Gas
Ekofisk	192	125	404	150
Frigg		127		112
Statfjord	341	40	535	54

Source: The Norwegian Directorate of Petroleum, *Annual Report*, various years.

line shows how recoverable reserves depend on the initial rate of production, with $q^* = 0.1$ and $s = 0.1$. The thin lines show how the initial rate of production depends on the investment cost per unit q_0. Optimum recoverable reserves are less than the maximum attainable, either because high investment costs make it undesirable to drill the necessary number of wells or because ''overexploitation'' is justified in order to get the incomes more quickly. We can also note the ambiguous effect of the rate of interest; at low cost a higher interest rate implies a higher initial production rate, in order to get the incomes more quickly.

Finally, it must be recognized that total reserves and the production rates of individual wells are far from being known with certainty, particularly before production starts and experience has provided additional information. Recoverable reserves and production rates depend on geophysical conditions such as permeability of oil-bearing rocks, impermeable layers that create barriers against the flow of oil inside the reservoir, and so forth. Needless to say, none of this is known with much certainty but the properties of the reservoirs are revealed through actual production. Technological improvements such as horizontal drilling may also increase the recoverable reserves. The figures for three Norwegian oil and gas fields in Table 4.2 illustrate the revision of reserves over time as experience has accumulated and technology progressed.

THE COMMON PROPERTY PROBLEM

The fact that different oil wells ''compete'' in extracting the oil introduces a problem known as the common property problem. The oil in a given reservoir is to some extent common for all the wells drilled into the reservoir. It would be an entirely common resource if it would flow without friction throughout the entire reservoir. This is not the case, but it is fugitive enough that the common element is a strong one, in any case for neighboring wells.

This is important because access rights to oil reservoirs are determined by access rights to the land above. In the United States ownership of land entails access to the mineral resources underneath. On the continental shelf outside the old three-mile limit the access rights to oil and other minerals belong to the

Table 4.3
The Norwegian Shares (Percent) of Three North Sea Fields in 1985 and 1995

	1985	1995
Statfjord	84.09322	85.46869
Frigg	60.82	60.82
Murchison	25.06	22.2

Source: The Norwegian Oil Directorate.

federal government, which leases these rights to private companies. Similar arrangements apply to the British and Norwegian continental shelves; licenses to oil prospecting and extraction are allocated by the governments of the United Kingdom and Norway, respectively.

If the ownership of land above an oil reservoir is fragmented, there are strong incentives to drill wells for the purpose of "stealing oil from one's neighbor" rather than optimizing the overall extraction from the reservoir. One additional well may add little value to the total extraction and may even make it less, as already explained, but for the owner of a small tract of land the well may be highly profitable, as it may allow him to get possession of some oil that otherwise would be extracted by his neighbor. This has been a major problem in Texas and Oklahoma where ownership of land above oil fields is often very fragmented. In offshore fields this is less of a problem, because the licensed tracts are large and there are typically few operators involved, and often just one. On the Norwegian shelf extraction licenses are not finalized until there is a plan in place for unitization of fields that are shared by two or more operators. Unitization means that the revenues from a reservoir being shared by two or more operators are shared in a given proportion, so that no one can increase his share of the reserves by accelerating production. Such rules govern, for example, the extraction from reservoirs that cross the Norwegian-British border in the North Sea. The Norwegian share of three such fields in 1985 and 1995 was as shown in Table 4.3. Two things are noteworthy. First, the shares change over time, as more is being learned about the reservoir, in particular how much oil there is on each side of the border. Second, the shares for Statfjord contain no less than five digits to the right of the decimal point, reflecting the fact that there are large sums of money involved.

To illustrate the waste that may arise because of the common property problem we can employ the simple oil tank model in Case I above. In this model oil flows without friction throughout the entire reservoir and could all be extracted through just one well; the only reason why there are more than one is the desire to get incomes early rather than late. The present value of the profit from the reservoir was given by Equation (4.14):

$$V = -cq_0 + pq_0(1 - e^{-rT})/r. \tag{4.14}$$

The decision variable, it will be recalled, was the production rate q_0, which is related to the number of wells drilled.

Now suppose that the ownership of land above the reservoir is fragmented. If all wells are drilled simultaneously they will all extract an equal amount of oil but less each, since there is a given amount of reserves to be extracted. As long as there is a positive net profit obtained by each well there will be incentives for each individual operator to drill an additional well. If there were few operators there might be scope for an explicit or implicit understanding among them not to drill too many wells since they are all made worse off by that process, but such understandings become less likely as the number of participants increases.

The initial rate of production arising when oil reserves are common property for many participants will therefore be given by setting $V = 0$, which entails

$$-c + p(1 - e^{-rT})/r = 0 \tag{4.23}$$

$$T = -\ln(1 - rc/p)/r, \tag{4.24}$$

where q_0 can be found by noting that the production rate will be constant in this case and total production will be equal to the initial reserves:

$$q_0 T = Q. \tag{4.25}$$

Table 4.4 compares the common property solution with maximization of present value. The production rate in the examples in Table 4.4 is two to five times

Table 4.4
Comparison of Maximizing Present Value and Rent Dissipation from a Simple Oil Tank Reservoir Model

	q_0	T	V
		c = 4	
Max. V	0.073	13.76	0.254
V = 0	0.196	5.11	0
		c = 1	
Max. V	0.188	5.32	0.588
V = 0	0.949	1.054	0

Note: Parameter values: $r = 0.1$ and $p = 1$.

greater under common property than needed to maximize the present value of profits, and the time until exhaustion is correspondingly shorter. The attainable profit is eaten up by unnecessary drilling costs. This is typical of common property resources, of which fish are perhaps the prime example. Fish swim around in the sea, much like the fictitious oil in this example, and all it takes to catch them if there are no access restrictions on fishing is to get a boat and go out to sea. The number of those who have access to fish, in the absence of any restrictions, is therefore high, and the attainable profits disappear in the form of costs for unnecessary fishing boats.

Whether the outcome for common property oil reserves in the real world will be better or worse than in this example is difficult to say. When recoverable reserves depend on the initial production rate, the outcome could be worse than indicated here. In addition to the loss implied by the drilling of unnecessary wells, some of the recoverable reserves might get lost permanently because of a too rapid fall in reservoir pressure. Examples of this are cited, for example, in Libecap (1989).

NOTES

1. The length of the plateau phase may be determined by the capacity of the processing equipment. Less than the theoretical maximum will be produced initially because of limited capacity to process, and the plateau phase becomes longer. This is particularly true of offshore installations that are constrained by space.

2. See, for example, McCray (1975), Chapter 11.

3. On discounting in continuous time, see footnote 8 to Chapter 3.

4. It is readily checked that $d^2V/dq_0^2 < 0$, $dV/dq_0 = 0$ is sufficient.

5. Brent Spar was an installation used by Shell to extract oil from underneath the North Sea, the usefulness of which had come to an end. Shell and the British government had come to the conclusion that the best way to get rid of it was to sink it into the Atlantic at great depth. Greenpeace, the environmental pressure group, mounted a big media campaign, arguing that this was not an environmentally sound procedure, and grossly exaggerated the amount of hazardous wastes onboard the installation. A public outcry followed, including, inter alia, incendiary bombs thrown against a Shell gas station in Germany. Shell backed off, after securing a temporary repository for the installation in a Norwegian fjord where it is still floating at the time of writing. It has, in fact, become a minor tourist attraction. The future fate of Brent Spar is still under study, involving considerable expenditure. For an update on the Brent Spar saga, see *Petroleum Economist* (November 1997); 18–19.

Chapter 5

Theories of Price Formation for Petroleum—Petroleum Rents

INTERTEMPORALLY OPTIMAL PRICES

Petroleum is a mineral resource in a fixed supply of an unknown magnitude. The fact that the supply is given would seem to have important implications for the price of this mineral and how it evolves over time. The second fact, that the supply is of an unknown magnitude, together with a third fact that it costs a lot to find out about it, complicates the picture considerably.

Consider first the case of full certainty; that is, pretend that we know exactly how much there is and that we also know other pertinent things about the future. The fact that the supply of petroleum is finite means that our present extraction must be done with a view toward the future. Presumably we would not like to run out of oil prematurely and unexpectedly. On the other hand, we would not like to be left with any of it at some future date when for whatever reason it might become useless.

The intertemporal allocation of a finite resource is, in a market economy, governed by prices. The price of oil must evolve over time in such a way that too prodigious use in any period is discouraged by a suitably high price. To focus on the intertemporal aspect implied by the finiteness of supply we shall employ a simple diagram. Let time be divided into two periods. The width of the open box in Figure 5.1 symbolizes the fixed and known supply of petroleum. The amount used in period 1 is measured from the left side and the amount used in period 2 from the right side, so that any point along the bottom line shows a possible distribution of the total supply between the two periods. Exactly how the supply should be allocated between the periods will depend on the usefulness of oil consumed in the two periods. In all probability the usefulness of an additional barrel of oil will become less and less the more we use.

Figure 5.1
Optimal Division of a Given Stock of a Nonrenewable Resource between "Now"
and "Later"

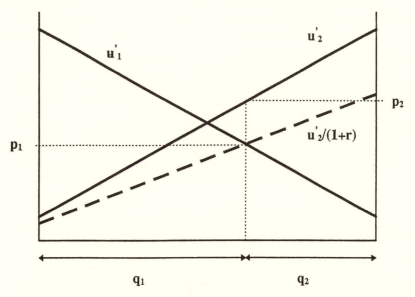

Note: The line u'_1 shows the marginal utility of the resource "now" and the line u'_2 the marginal
utility "later," and $u'_2/(1 + r)$ is the value "now" of the marginal utility "later." Optimum
allocation between now and later is given by the intersection of the lines u'_1 and $u'_2/(1 + r)$,
with q_1 being used "now" and q_2 "later." The prices accomplishing this allocation are p_1 and
p_2, respectively.

The downward-sloping lines in Figure 5.1 are meant to show this falling marginal utility (u') of oil in each period.

The best possible allocation of the oil between the two periods would appear to be where the lines u'_1 and u'_2 intersect. At this point we would be indifferent between burning up one additional barrel of oil in period 1 and period 2; in a situation like that there would be nothing to be gained from changing the allocation between periods. But things are not quite so straightforward. We usually prefer things now rather than later; everything else being equal, we would prefer a given sum of money now rather than a year from now, even if we could be sure that it did not lose any of its purchasing power in the meantime. Because of this impatience we would have to discount the marginal utility of oil in period 2 to make it comparable to the marginal utility in period 1. The dashed line in Figure 5.1 shows the discounted marginal utility in period 2, $u'_2/(1 + r)$, r being the rate of discount. Equalizing marginal utilities across periods, we find that the best allocation must satisfy the criterion

$$u'_1 = u'_2/(1 + r). \tag{5.1}$$

Now, in a market economy the allocation of goods over time is determined by prices. If a commodity costs twice as much as another commodity, the marginal utility of the first commodity must be twice as high as that of the second, for otherwise a rational consumer would change the proportion in which he consumes the two commodities. We can apply a similar reasoning for the allocation of oil over time. Therefore, the ratio of oil prices in two adjacent periods would in an optimal allocation be equal to the ratio of marginal utilities

$$p_2/p_1 = u'_2/u'_1. \qquad (5.2)$$

Substituting this into (5.1) we have

$$(p_2 - p_1)/p_1 \equiv \Delta p/p = r; \qquad (5.3)$$

that is, the price must rise over time at a rate equal to the rate of discount.

This simple example has several interesting implications. First, we have said nothing about the cost of production; we have proceeded as if it were zero. Nevertheless there is a positive price. This price is due simply and solely to the finiteness of the resource. The implication is that resources in finite supply have a value over and above what it costs to produce them, owing to their scarcity.

The role of the price is to ration the use of the resource over time. With discounting of the future we want to consume a bit more in period 1 than in period 2. Therefore, the price must be lower in period 1 than in period 2 to facilitate greater use in period 1. The price must rise at a rate equal to the rate of discount in order to bring about the best possible intertemporal allocation. But what are the forces on the supply side that ensure that the resource will be forthcoming in exactly the proportion found to be appropriate by looking at the use or the demand side? As already noted, there are no production costs; no compensation is needed to bring forward any supply. But why should an owner of the resource be willing to supply the resource in exactly the right amounts? The reason why a resource owner would not dump everything on the market in period 1 is the capital gain he can make by waiting until period 2. If he extracts one unit now and invests the money he gets from selling it in the capital market at r rate of interest, he will earn a return equal to rp_1. If he supplies the unit in period 2 it will have appreciated in price, so he will make a return equal to $p_2 - p_1$ simply by waiting until period 2. Suppliers will be indifferent between supplying in the two periods only when these returns are equal, which means that (5.3) is satisfied. Hence (5.3) must be satisfied if supply and demand are to be equal in both periods. The rule that the price of a finite resource must rise at a rate equal to the rate of interest is called the r percent rule or the Hotelling rule, after the American economist Harold Hotelling, who in 1938 published a paper on resource pricing that now has become a classic.

INTERTEMPORALLY OPTIMAL PRICES:
A FORMAL ANALYSIS

The division of time into "now" and "later" is a crude one but sufficient for illustrating some salient features of a price system consistent with an intertemporally optimal allocation. But what happens when "later" is over? How late is "late"?

One possible use of the intertemporal price theory developed above is the following. Let us pretend we can make a reasonable guess at how much oil there is left in the ground. Modern life as we know it need not come to an end when we run out of oil. There are different substitutes for oil in different uses; it is possible to produce gasoline from coal, electricity from nuclear reactors, and so forth. It is, however, much more expensive to produce gasoline and electricity in this way than from oil. So, if the price of oil were higher than what corresponds to this cost, people would start producing gasoline from coal and electricity from nuclear power plants to a greater extent than is being done already. Conversely, immediately before we run out of oil it could be sold at a price just below the price that induces people to stop using oil.

The highest price at which oil could be sold without becoming uncompetitive has come to be known as the "backstop price," a term borrowed from baseball. Hence we know where the price ends up immediately before we run out of oil. From the r percent rule we know that the price must rise by r percent per year (r could, of course, change over time). From this we can back-calculate the price, from the end to the present. The only unknown, then, is the length of time from the present until we run out of oil. But knowing the price in any period means knowing the quantity demanded, provided other factors upon which demand depends can be predicted. By adding together the demand in all periods and setting it equal to the amount of oil left in the ground at the present time we can find the length of time from now until we run out of oil.

It helps to formalize these musings. Let the backstop price be denoted by p^* and the time at which we run out of oil by T. Because of the r percent rule, the price at any time t will be

$$p_t = p^* e^{r(t-T)} \tag{5.4}$$

The time derivative is $dp/dt = rp$, as required by the r percent rule.

The quantity of oil demanded at time t, assuming that all factors other than price will remain stable over time, will depend on the price only. For the sake of getting an analytical solution, assume a demand function with a constant price elasticity b:

$$q_t = A p_t^{-b} = A(p^*)^{-b} e^{rb(T-t)} \tag{5.5}$$

At time 0, the quantity left of oil is S_0. Total extraction from then on until we run out of oil is the sum of q's from 0 to T, or in continuous time the integral of q from 0 to T:

$$\int_0^T q_t dt = \frac{A}{br(p^*)^b}(e^{brT} - 1) = S_0 \tag{5.6}$$

from which we can find T as

$$T = \frac{1}{br}ln + (1 + \frac{br(p^*)^bS_0}{A}) \tag{5.7}$$

where T is the time we run out of oil, or the length of the time period over which it is optimal to plan to extract the oil.

No one knows, of course, how much oil there is left, but anyone who wants to plan for the future will have to make forecasts about things to come. By making the best possible guesses about the backstop price, interest rates, amount of oil left, and future demand for oil it is possible to use this approach to forecast future prices. The reliability of such forecasts will depend on how appropriate a description of the workings of the oil market this approach represents, a point to which we will return below.

Figure 5.2 shows the development paths of price and the extraction rate (q) over time (Equations [5.4] and [5.5]). The figure also shows the effect of raising the discount rate. A higher discount rate means that the price will increase more rapidly; that is, the price path will become steeper, but it must end at the backstop price p^*. The price cannot, however, reach the backstop price at the same time point as before. If this were so, the price would be lower at all time points $0 \leq t < T$, which means that the rate of extraction would be greater, and the total extraction over the period T would be greater than the resources available at time 0. Therefore, the price path for a higher discount rate must intersect the price path for a lower discount rate, as shown in Figure 5.2. A higher discount rate means that the price will be lower initially and the rate of extraction greater, and the total period of extraction will become shorter (T will become smaller). This can be readily seen by assigning numbers to the parameters in Equations (5.4) through (5.7).

Increasing the amount of oil available at time 0 would lower the price at any future point in time and lengthen the extraction period (raise T); since there is more available, more must be extracted. Raising the backstop price has the opposite effect; it pulls the price path upward, so less will be extracted at any future date, but the extraction period must become longer, for otherwise all the deposits available at time 0 would not be extracted.

Nothing has been said so far about costs of extraction. These costs do affect the above conclusions to some extent, but the r percent rule still holds for the

Figure 5.2
Optimum Price and Extraction Paths for a Nonrenewable Resource of a Given Size, with a Constant Elasticity of Demand Equation that Remains Fixed over Time

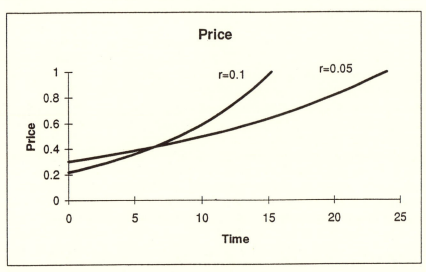

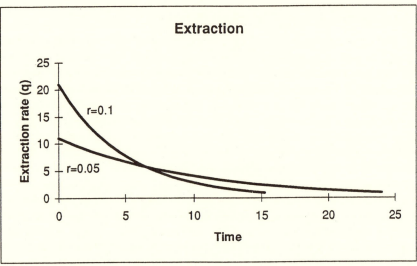

Note: Demand equation: $q_t = A(p^*)^{-b}e^{rb(T-t)}$, with $A = 1$, $p^* = 1$, $b = 2$, and $S_0 = 100$.

net price rather than the market price. Denoting unit cost of extraction by c, the criterion for intertemporal optimality and competitive equilibrium is

$$(p - c)r = d(p - c)/dt \tag{5.8}$$

The term on the left side of the equation shows the return from extracting one unit of oil and investing it in the capital market at r percent rate of interest while the term on the right shows the gain from delaying extraction (the time derivative of the net price). The gain from deferring extraction may come both from a rising price and from a lesser rise in unit cost of extraction than otherwise would occur.

If the unit cost of extraction is not constant it will, in a world of perfect foresight, rise over time; it will make sense to deplete the cheapest resource deposits before the more expensive ones. Ultimately the unit cost of extraction may catch up with the price, in which case extraction comes to a halt because the marginal utility of the resource is not sufficient to justify the marginal cost of digging it up. The size of the resource deposit will in that case be determined by economic factors, i.e., the marginal cost of extraction compared to the marginal utility of the resource. This is in fact the typical case for mineral deposits; oil wells are abandoned long before they have been physically emptied; often as little as 20 percent is extracted from oil wells, and it would certainly be cheaper to substitute coal, tar sands, and nuclear power for oil as a source of energy than to squeeze the very last drop out of abandoned oil wells.

INTERTEMPORAL PRICES AND MARKET ORGANIZATION

Competitive Versus Monopolized Markets

We have established that an intertemporally optimal allocation of a finite resource will require that its net price rise over time at a rate equal to the rate of interest. Precisely this will occur in a competitive equilibrium. This is so because if (5.8) does not hold, the owners of oil deposits will have incentives to accelerate or to withhold production. If the left-hand side of (5.8) were greater than the right-hand side, it would be profitable for resource owners to accelerate extraction and to invest the proceeds at the going rate of interest (r). If, on the other hand, the right-hand side of (5.8) is greater than the left-hand side, it would be better to let the resource appreciate in value while lying in the ground. Therefore, the only price path that is compatible with equilibrium implies that owners of oil deposits are indifferent between these two alternatives so that (5.8) holds. Since the acceleration of production implies that the market price will fall, whereas withholding production implies that it will rise, it is possible to conceive of the equality in (5.8) as being brought about by market forces.

This conclusion rests crucially on the assumption that the owners of oil deposits are so many that they do not coordinate their actions to manipulate the price; each takes the price as given at any point in time. This would not be true, of course, in a world where there is just one owner of oil deposits, or more to the point, where there are just a few. Let us look at the case with just one owner first, because it is simple and in clear contrast with the competitive case. If there is a backstop price, the sole owner could withhold his production as necessary to charge the backstop price as long as there are any deposits left. After all, this is the highest price that can be charged, and it would appear to be in the interest of the resource owner to charge that price and no less. On the other hand, the sole owner might be impatient and want his incomes earlier rather than later. To accelerate the flow of income, he would have to lower the price.

Suppose the price charged by the sole owner is lower than the backstop price. Maximum profit for the sole owner would obtain when he is indifferent between extracting one unit of oil immediately or deferring extraction to a later date. Since the sole owner affects the market price by his rate of extraction, the version of (5.8) that is relevant in this context is

$$(mr - c)r = d(mr - c)/dt \tag{5.8'}$$

where mr denotes the marginal revenue from extraction. Since marginal revenue is related to price by the formula

$$mr = p(1 - 1/e), \tag{5.9}$$

where e is the price elasticity of demand, the rise in marginal revenue implied by (5.8') implies a rise in price as well, but at the same rate only if the elasticity of demand is constant.

This last remark alerts us to a possible difference between monopolizing the market for a nonrenewable resource such as oil and monopolizing the market for an ordinary commodity produced by the application of ordinary inputs, independently of any nonrenewable inputs. If the price elasticity of demand is constant, both the marginal revenue and price of a nonrenewable resource will rise at the same rate and one that is equal to the rate of discount. This is in fact the same rule as holds in a competitive market. Assume, for the sake of the argument, that the backstop price is infinitely high so that it will be attained only asymptotically and the price will rise continuously at the rate of interest. The price path would then be the same under monopoly as in a competitive market, implying that the sole owner would always supply the same amount as the competitive market. This is in marked contrast to the conventional monopoly theory, which says that a monopolist will always supply less than the competitive market. But on closer look the result is not, perhaps, so surprising. What degrees of freedom does a resource monopolist have, compared to a competitive market? He can choose a different time profile for his supply and manipulate the price, but he has nothing to gain by withholding permanently any of his deposits. It is not obvious that the monopolist has anything to gain from choosing a time profile of extraction different from the one that would obtain in a competitive market. This is very different from monopolizing the market for an ordinary commodity that is produced through a flow of inputs and not by depleting a finite stock of resource deposits. In the first case it pays for the monopolist to choose a less intense flow of inputs than would obtain in a competitive market and so limit permanently a sustained flow of output to get a better price.

Nevertheless, there is a presumption, as we shall see, that an oil monopolist would in fact choose to supply less initially than a competitive market would do and so stretch the supply of oil a little further. This happens when there is an upper limit to the price (the backstop price) and cost of extraction to be reckoned with, as will be demonstrated below.

With a backstop price, the monopolist's extraction of the oil can be divided into two phases, one phase during which the marginal revenue is rising at the rate of discount and another phase during which the price is equal to the backstop price. The present value of the entire extraction, at a constant rate of interest, would be

$$PV = \int_0^{T_1} f(p_t)p_t e^{-rt}dt + p^* f(p^*) \int_{T_1}^{T_2} e^{-rt}dt \qquad (5.10)$$

where $f(p_t)$ is the demand for oil at time point t. The demand for oil depends, of course, on many other factors than its price, but in order to focus on the problem of intertemporal optimization we shall pretend that these factors are always the same and that they do not change over time. We also ignore extraction costs. In the first phase, from time 0 to T_1, the price will change in such a way that the marginal revenue rises at the rate r, as already explained, while in the second phase, from T_1 to T_2, the monopolist charges the backstop price and always extracts the same quantity.

How long are the two phases? To illustrate this more clearly we shall use the same demand function as in (5.5) above, $q = Ap^{-b}$. The price elasticity of demand is $e = b$, and the marginal revenue is $mr = p(1 - 1/e) = p(1 - 1/b)$. Since the price elasticity of demand is constant, price will increase at the same rate as the marginal revenue, which rises at the rate r during the first phase of extraction. Hence the price at any point in the interval 0 to T_1 will be

$$p_t = p^* e^{r(t-T_1)} \qquad (5.10)$$

Substituting this into (5.10) we get

$$PV = \frac{Ap^{*1-b}}{r}e^{-rT_1}\left[\frac{1}{b}(e^{rbT_1} - 1) + (1 - e^{-r(T_2-T_1)})\right] \qquad (5.10')$$

Since the monopolist has fixed deposits and will extract all of them in due course, he has only one degree of freedom, namely how much to extract during phase 1. This will determine how much he starts with during phase 2, and the length of both phases. Hence, both T_1, the length of phase 1, and $T_2 - T_1$, the length of phase 2, can be seen as dependent on S_1, the amount left behind after phase 1. To find the optimal length of phase 1 we can take the derivative of PV with respect to S_1 and set this equal to zero. This gives

$$\left[\left(\frac{1}{b} - 1\right)(1 - e^{rbT_1}) + e^{-r(T_2 - T_1)}\right]\frac{dT_1}{dS_1} + e^{-r(T_2 - T_1)}\frac{d(T_2 - T_1)}{dS_1} = 0 \qquad (5.12)$$

It remains to evaluate the two derivatives dT_1/dS_1 and $d(T_2 - T_1)/dS_1$. To do so, note that

$$S_0 - S_1 = \int_0^{T_1} q_t dt = \frac{Ap^{*-b}}{rb}(e^{rbT_1} - 1) \qquad (5.13)$$

so that

$$T_1 = \frac{1}{rb}ln\left(\frac{rbp^{*b}}{A}(S_0 - S_1) + 1\right) \qquad (5.14a)$$

while

$$T_2 - T_1 = \frac{S_1}{Ap^{*-b}} \qquad (5.14b)$$

From this we get

$$\frac{dT_1}{dS_1} = -\frac{p^{*b}}{rbp^{*b}(S_0 - S_1) + A} \qquad (5.15a)$$

$$\frac{d(T_2 - T_1)}{dS_1} = \frac{1}{Ap^{*-b}} \qquad (5.15b)$$

Inserting (5.15a) and (5.15b) into (5.12) we can find T_1, for any given initial reserves and other parameters.

Figure 5.3 compares the price path and extraction rate obtaining under monopoly versus perfect competition. The parameters in the above model have been set as follows: $p^* = 1$, $A = 1$, $b = 2$, $r = 0.05$, and $S_0 = 100$. The two phases in the monopoly solution emerge very clearly. The competitive price path is below the monopoly path and hence the extraction rate is greater under competition; in the competitive solution the reserves are exhausted in less than twenty-five years while under monopoly they last for over thirty-five years. Hence the monopoly conserves resources to a greater degree than would occur in a competitive market, owing to the backstop price and the existence of phase 2 in the monopoly solution. A similar tendency to resource conservation would

Figure 5.3
**Price and Extraction Profiles under Monopoly (Thick Lines) versus Competition
(Thin Lines), with a Constant Elasticity of Demand Equation that Remains Fixed
over Time**

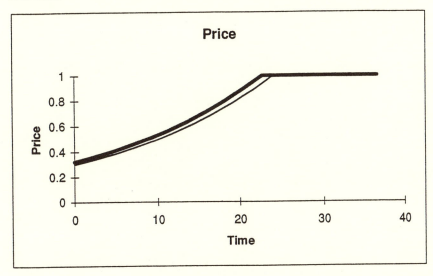

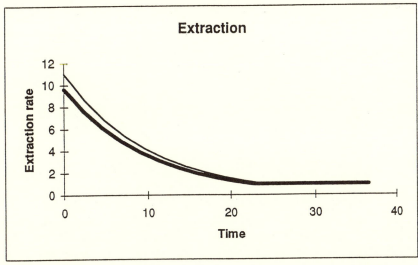

Note: Demand equation: $q_t = A(p^*)^{-b}e^{rb(T-t)}$, with $A = 1$, $p^* = 1$, $b = 2$, $r = 0.05$, and $S_0 = 100$.

result from a positive extraction cost. It can be verified that phase 1 shrinks as the total deposits become smaller, but it would be necessary to reduce the total reserves to about one unit to make phase 1 disappear totally.

Imperfect Competition

In the real world markets are characterized neither by perfect competition nor by monopoly but by something in between. This is certainly true of the oil market. For many years OPEC exerted a strong influence on the oil price but it has seen its power diminish in recent years. In OPEC's heyday it was common to regard it as a price leader in the market for oil.

It is possible to deal with price leadership within the intertemporal price theory that we have been discussing. There exists considerable literature on this, but since the intertemporal price theory of the kind developed above does not appear to explain the development of oil prices very well we shall not go very deep into this literature, which is mathematically demanding.[1] Here we shall try to highlight the issue with a very simple approach. Suppose a part of total reserves is owned by many competitive firms, which take the market price as being given and outside their sphere of influence. For these "fringe producers," as they are often called, the price must rise over time at the rate of interest (we continue to ignore extraction costs). The price leader must take this into account. Using the same specification of the demand function as above makes the exposition very simple, because the price would rise by the rate r in phase 1 under both monopoly and perfect competition. The price leader and a competitive firm would obtain the same profit for any unit of oil sold, and the present value of that profit, during phase 1, would be the same at all points in time. Denoting the total present value of oil production by $PV = V_1 + V_2$, where the subscripts refer to the two phases, respectively, the price leader would obtain a fraction s of V_1, where

$$s = (S_{0L} - S_{1L})/(S_{0L} - S_{1L} + S_{0F}) \qquad (5.16)$$

and L and F denote the price leader and the competitive fringe, respectively, and S_0 and S_1 are the reserves at the beginning and end of phase 1 (the competitive fringe would, of course, exhaust all its reserves during phase 1, since the price is constant during phase 2).

Hence the present value of the leader's extraction would be

$$PV_L = sV_1 + V_2 \qquad (5.17)$$

where V_1 and V_2 refer to the two terms in Equation (5.10) above. As in the monopoly example the leader selects the reserves to leave behind at the end of phase 1. The optimality condition is

Figure 5.4
Dynamically Inconsistent Price Path with a Leader Controlling 80 Percent of Reserves at Time 0, with a Constant Elasticity of Demand Equation that Remains Fixed over Time

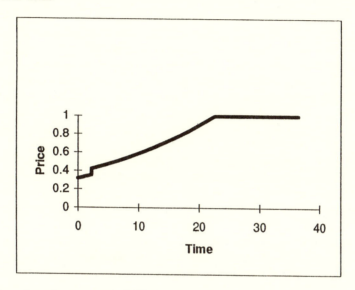

Note: Demand equation: $q_t = A(p^*)^{-b}e^{rb(T-t)}$, with $A = 1$, $p^* = 1$, $b = 2$, $r = 0.05$, and $S_0 = 100$. The competitive fringe is assumed to extract its reserves first.

$$-\frac{S_{OF}}{(S_{OL} - S_{1L} + S_{OF})^2}V_1 + s\frac{dV_1}{dS_{1L}} + \frac{dV_2}{dS_{1L}} = 0 \qquad (5.18)$$

where dV_1/dS_{1L} and dV_2/dS_{1L} are as in (5.12).

Figure 5.4 shows the price path resulting from a price leader optimizing his extraction program. The price leader is assumed to control 80 of 100 units of reserves at time zero. Making up his plan at time 0, the price leader finds that he wants to extract about 69 units during phase 1 and leave behind about 11 units for phase 2. The fringe suppliers will probably have exhausted their supplies long before the end of phase 1, so we may ask what happens then? Would that affect the solution found to be optimal at time zero, when the fringe suppliers were present? The answer is yes, it would. Suppose the fringe exhausts its supplies first. Then, when the fringe supplies have disappeared, which happens at about time 2.5, the leader recalculates his program with 80 units of deposits and no fringe, so he is in fact a monopolist. He now finds that he wants to extract only 66 units during phase 1 and leave behind 14 units for phase 2. This makes it necessary to reduce the demand during phase 1 and to adjust the price upward, as shown by the jump of the price path in Figure 5.4.

This illustrates a problem known as dynamic inconsistency. That is, when circumstances have changed the planner wants to change his plan, even if these circumstances are a consequence of the plan itself and could be foreseen when the plan was made at time 0. The question arises whether such a plan is credible. Presumably the competitive suppliers will be able to see through this and adjust accordingly; for if this plan were followed it would be profitable for the fringe suppliers to withhold their supplies until the price path jumps upward, as the capital gains from price appreciation of unextracted deposits would outweigh the profits of extraction. With perfect foresight on behalf of all agents the only credible outcome is, perhaps, that the price path be smooth and rising at a rate equal to the rate of interest.

OIL PRICES AND COSTS OF DISCOVERY

If we look at the development of the oil price over time, we do not see much of a verification of the r percent rule. Prices have not risen smoothly at a constant rate, or even a variable rate. For many years before 1970 the price declined. Then came the two price hikes in the 1970s, and a steep fall in 1986. Since then the prices have varied, with a downward trend if anything.

It is true that this development in and of itself does not contradict the r percent rule. This rule is based on the assumption of known reserves. In reality, reserves are not known but are discovered gradually at a considerable cost. Every now and then huge reserves are discovered and sometimes with lower costs of extraction than existing reserves. Oil price forecasters would revise their calculations in the light of such findings; a large increase in reserves and a fall in costs of extraction would push down the calculated price path. This would lead to a fall in the price, not because the r percent rule is wrong and irrelevant but because new evidence has emerged, changing the parameters in the models applying the rule. Furthermore the oil market is anything but competitive; as already discussed, we have seen OPEC's influence wax and wane, and in response to that the price has risen and then fallen; the r percent rule might nevertheless have been perfectly valid under a stable market regime.

Still the r percent rule seems more than a little out of place. Few, if any, oil companies seem to plan their extraction policies on the basis of the r percent rule. There is an alternative theory that seems more pertinent to what companies actually do, and has in fact implications that are not very different from those of the r percent rule. This theory does not take the finiteness of oil reserves as a point of departure. On the contrary, it regards oil just as any other commodity available in any supply we would care to have, but at a price.[2] It takes a long and expensive process of prospecting, exploring, and development to bring supplies from an oil field on stream. Oil companies like to have inventories of oil reserves for ten to twenty years of production and are therefore engaged in prospecting for oil all over the world. This activity has been so successful that the known oil reserves of the world have kept pace with extraction; for the last half century the world has had oil reserves for thirty to forty years of production despite a tenfold increase in the annual consumption of oil. Again and again somebody takes the reserve figures and predicts that we will run out of oil in thirty to forty years. Like all other doomsday predictors, they have been proved wrong again and again.

What does this imply for oil prices? At any given point in time there is a multitude of oil fields in production, under development, or for which decisions whether or not to develop are pending. Some will not be profitable at the prevailing price and will be abandoned or not put on stream if not already developed. If oil prices are "low" and expected to remain so, prospecting and development of new fields will suffer. If current demand cannot be met by production from existing fields, the price will rise, making it profitable to put previously unprofitable fields on stream and intensify prospecting for new finds. Hence, in a long-run perspective, the price of oil must cover both ordinary production costs and the costs of finding new deposits. In this scenario there will be a price premium over and above ordinary costs of production that does not reflect the finiteness of oil reserves but the cost of finding new ones.

The implication for the price path for oil is somewhat similar to the r percent rule. Since oil companies can be expected to explore and develop the least costly areas first, we can expect the minimum finding cost to drift upward over time. But every once in a while nature will spring a surprise. New areas that previously were thought uninteresting will turn out to have plenty of oil and possibly allow inexpensive extraction. The North Sea was for a long time believed to be uninteresting; when prospecting for oil underneath the North Sea was first put on the agenda some Norwegian geologists regarded the probability of finding oil deposits there as insignificant. Such events will lower the minimum finding costs and exert a downward pressure on the price of oil. Hence, as with the r percent rule, we would observe rising prices over time, a development occasionally reversed by surprise finds. Technological progress will also exert a downward pressure on prices; in fact it is entirely possible that it will be important enough to reverse the tendency to rising prices that would otherwise follow from successively exploiting less and less lucrative areas.

PETROLEUM RENTS

Oil fields come in many different sizes and shapes and are to be found in highly different locations: in the desert of Arabia, the Amazon jungle, the Siberian tundra, and underneath the North Sea. The differences in the cost of production are huge, from something like one or two US dollars per barrel in Saudi Arabia to 10 dollars or more in the North Sea. One may wonder why oil is being produced simultaneously from all these fields and why the market is not flooded by oil from the low-cost fields, with low-cost producers undercutting the high-cost producers and putting them out of business. The answer is that production from the low-cost fields would hardly suffice, and certainly not for long, to satisfy world demand, particularly not at a lower price. Furthermore the low-cost producers are not interested in selling their oil at bargain prices; despite its weakness OPEC still tries to restrain the production from its member states, with some success. Under any circumstances there would always be some cost differences between fields; the production costs from a field are never known exactly until the field has been put on stream, as the geological structure of any given field, the amount of oil in place, and its flow rate can never be known with full certainty.

These cost differences imply that there is a substantial profit to be earned on the cheapest fields. This profit is analogous to land rent; while the marginal piece of agricultural land just breaks even, the more productive and better located land will earn a profit. Hence in the oil industry there will be a differential rent to be earned on the more productive and advantageously located oil fields.

In addition to the differential rent there is market power rent to be reckoned with. As discussed earlier, OPEC exerted a huge influence on prices in its heyday by producing less than it could have done and probably still does so to some extent; there is no doubt that certain OPEC members could flood the market with cheap oil and drive the price down to well below 10 dollars a barrel but whether it would be in their interest to do so is another matter; their total revenues would most likely fall and they would be increasing the need for themselves to find new reserves. To the extent that OPEC restrains its production the price will be raised over and above the level necessary to cover the minimum finding cost, and a rent will emerge that is due to the exercise of this market power. Hence there are two kinds of rent in oil production, differential rent and cartel rent.

OPEC AS A PRICE LEADER

The theory of price leadership can explain how the market price can exceed the marginal cost of production, including the minimum finding cost. Figure 5.5 illustrates this theory in a simple way. There is a competitive fringe with a rising marginal cost of production, MC_f, including minimum finding cost. The fringe will supply the quantity at which the price is equal to its marginal cost. Subtracting the fringe supply from the market demand (D_m), we arrive at the residual demand facing the dominant supplier, denoted D_d. The marginal revenue corresponding to this residual demand curve is MR_d.

Now suppose, for simplicity, that the marginal cost of the market leader is constant and equal to MC_d. His profit-maximizing quantity will be where his marginal revenue is equal to his marginal cost, at Q_d. Adding the quantity supplied by the fringe, Q_f, we arrive at a point on the market demand curve (Q_m) and see that the market price is well above the marginal cost of the leader, but equal to that of the fringe. Implicit in this reasoning is that the marginal cost of the leader is below that of the fringe, which is indeed reasonable for the case of OPEC versus the rest of the world.

Over time the marginal costs of the fringe and the leader would be expected

Figure 5.5
Optimal Price for a Dominant Firm in a Market with a Competitive Fringe

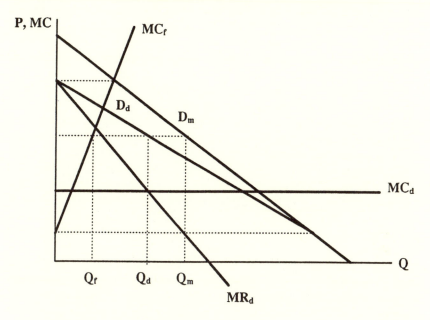

Note: Market demand is D_m, the marginal cost, and the supply curve of the competitive fringe, is MC_f, leaving D_d as the residual demand for the dominant firm. The marginal revenue for the dominant firm will be MR_d, and with the marginal cost of the dominant firm being MC_d, the optimum quantity supplied by the dominant firm will be Q_d, while Q_f will be supplied by the competitive fringe.

to drift upward due to exploration of less and less promising areas, but surprise discoveries and improvements in technology will counteract this, and the net effect may possibly be a lowering of the marginal cost. The marginal costs of the fringe and the leader may be differently affected by these developments, altering the balance of market power between the two.

This leader-follower model may throw some light on the developments in the oil markets since the first oil price hike in 1973–74. In the short term the demand for oil is very inelastic, as indicated by the curve D_{m1} in Figure 5.6 (the same as D_m in Figure 5.5). Suppose the prevailing price is equal to the marginal cost of the dominant supplier, MC_d. The dominant supplier realizes that his optimum price is much higher, or P_1, given the inelastic market demand, and raises his price accordingly. At this price he will supply the quantity Q_{d1}. In the longer term the demand for oil turns out to be much more sensitive to price, however, as indicated by the curve D_{m2}. If he holds on to the price P_1, the leader will see his sales shrink dramatically. Suppose that the fringe producers are already committed to producing the amount where their marginal cost is equal to the price

Figure 5.6
Optimal Price for a Dominant Firm in a Market with a Competitive Fringe,
before and after Demand Has Been Revealed To Be Elastic

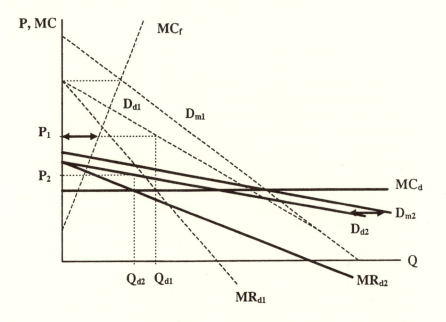

Note: The demand curve is originally believed to be D_{m1}, which is less elastic than D_{m2}. The optimum price will be believed to be P_1 (the same as in Figure 5.5) while the true optimum price is much lower. Setting the price equal to P_1 entices the competitive fringe to supply the quantity shown by the lines with arrows. Even if the price falls, the competitive fringe will continue to supply this quantity because of sunk costs. The residual demand for the dominant firm will therefore be D_{d2} (given by displacing D_{m2} horizontally by the quantity supplied by the fringe), and the optimum price will be P_2.

P_1, even if the price falls below that. The idea is that the fringe producers have already developed fields on the assumption that this price will prevail and would not shut them down even if the price falls, because operating costs are much lower than the development costs. The demand curve facing the dominant producer would then be D_{d2}, that is, the market demand shifted to the left by the amount produced by the fringe, as indicated by the line with arrow at the ends.

For this more elastic demand curve the optimum price for the leader would be P_2, which is lower than the leader would have thought originally if he believed in the inelastic demand curve D_{m1}, and his optimum quantity would be Q_{d2}. We see that the leader runs into short- to medium-term problems when he reverses his decision to raise prices. The marginal cost curve of the fringe was defined as including development and finding costs, implying that the higher price engineered by the leader led to development of new, high-cost fields and increased supply by the fringe. But reversing the price will not cause the supply of the fringe to decrease; development costs are sunk, literally, in the form of oil wells and their equipment, oil platforms at sea and installations on the sea bottom, equipment that is of little or no use for anything else than oil production. The fringe producers will continue to produce as long as they get their operating costs covered, even though the new and lower price will not justify the development of the high-cost fields.

Possibly the OPEC leaders believed in an inelastic demand for oil justifying the high price they engineered in the 1970s and early 1980s. But two things happened. The ever-growing demand for oil came to an abrupt halt, owing to the substitution of oil by other sources of energy offsetting the increase in demand due to economic growth. Second, new, high-cost areas such as the North Sea came on stream. The upshot of both of these was that OPEC's share of the market shrank dramatically. Production in the North Sea continued and in fact expanded after the fall in oil prices in 1986, not only because sunk costs are sunk but also because of impressive cost-cutting technological progress, stimulated in part by the fall in oil prices.

WHO GETS THE PETROLEUM RENT?

In the absence of taxes or fees, the rents from petroleum fields accrue to those who exploit these fields, and the rents will be higher, the lower the costs of production. As discussed in the next chapter, governments in petroleum-producing countries try to skim off a substantial part of these rents in various ways. In this section we shall be concerned with the division of the rents between petroleum-producing and -consuming countries. As discussed in Chapter 2, the highly nonuniform distribution of petroleum deposits divides the countries of the world into producers and consumers of petroleum. Some countries are, however, both major producers and consumers, such as Russia, the United Kingdom, and the United States.

In most countries oil products are subject to substantial and sometimes very high taxes. Table 5.1 shows the taxes as percent of the final market price for three countries and four products. These three countries represent high-tax, medium-tax, and low-tax regimes. In France the taxes on oil products are very high, but not much higher than in most other western European countries. In the United States these taxes are substantially lower than in Europe, and in Mexico these taxes are lower still. Another feature we can note from Table 5.1 is that the taxes on gasoline and diesel in France (which is typical of European countries) are much higher than the taxes on other products such as fuel oil.

The motivations for taxing oil products are several: purely fiscal, environmental (less local pollution and less emissions of greenhouse gases), and in some countries protection of competing domestic industries, coal in particular. It is possible, however, to view these taxes as devices to divert some of the petroleum

Table 5.1
Taxes of Oil Products as Percent of Final Market Price, 1996

	Gasoline	Auto diesel non-com- mercial	Light fuel oil households	Heavy fuel oil industry
France	79.1	70.5	40.1	13.3
United States	31.1	35.8	n.a.	n.a.
Mexico	13.0	13.0	n.a.	13.0

Source: International Energy Agency, *Energy Prices and Taxes* (Paris: OECD).

rents to the governments in petroleum-importing countries and away from petroleum-exporting countries. Intentionally or not, this diversion of the rents will inevitably result from the application of these taxes. Suppose the market of oil were competitive. A tax on oil products in oil-importing countries translates ultimately into a tax on oil imports, and the tax drives a wedge between the final market price (back-calculated from the prices of the final products) and the import price of crude oil. Figure 5.7 shows how the tax is borne by buyers and sellers, and how the total amount used falls as a result of the tax. The proportion in which the buyer and seller, respectively, bear the tax depends on the elasticity of demand and supply, as is dealt with in standard textbooks on economics.

In reality the oil market is not fully competitive; OPEC or enlightened self-interest alone has the effect that the biggest low-cost producers do not glut the market with oil to the extent they are probably capable of. Be that as it may, the important point is that taxes on oil products drive a wedge between the final market price and the import price. The oil producers would undoubtedly rather have restrained the supplies of oil by their own endeavor and cashed in on the higher price for their own benefit, without having any of it disappearing into the coffers of the governments in the oil-importing countries.

Since the steep fall in oil prices in 1986, the governments in oil-importing countries have raised their taxes on oil products and obtained an ever-increasing share of the oil rents. Figure 5.8 shows the development of crude oil import costs and energy prices for end-users in the OECD countries. Import prices more than doubled from 1978 to 1981, owing to the second oil price hike, and then fell steeply in 1986 and have remained relatively stable ever since. On the other hand, energy prices have remained relatively stable over the whole period, with a downward trend. This implies that taxes have increased and prevented the final prices of energy from falling in tandem with the price of crude oil imports.

The increase in taxes on final products has been fairly uniform since the early

Figure 5.7
The Effect of a Tax on Equilibrium Price (*P*) and Quantity Traded (*Q*) in a Competitive Market

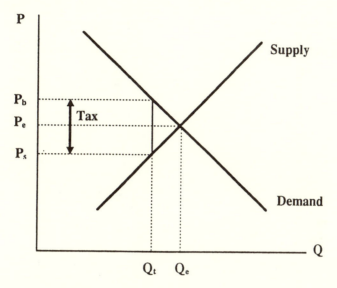

Note: In the absence of a tax the equilibrium price will be P_e and the quantity traded will be Q_e. The tax drives a wedge between the price paid by the buyer (P_b) and the price received by the seller (P_s), and the quantity traded will be reduced to Q_t.

to mid-1980s. Figure 5.9 shows the development of taxes on gasoline and automotive diesel in the United States and the United Kingdom and on light fuel oil for household consumption in the United Kingdom and Denmark. Taxes on gasoline and on automotive diesel have risen in both the United States and the United Kingdom, but remain substantially lower in the United States. Taxes on light fuel oil for household consumption rose steeply in Denmark in 1985 and have remained between 60 and 70 percent of the market price ever since, but in recent years taxes on light fuel oil have risen substantially in the United Kingdom as well. Denmark and the United Kingdom are at opposite ends in the European league of tax regimes for fuel oil.

Figure 5.10 shows the development of final prices and taxes of these products. In both the United States and the United Kingdom the taxes on gasoline have risen much more than the prices. The same is true of light fuel oil in Denmark and the United Kingdom; in Denmark there was a steep rise in the tax in 1985, as already noted, and in the United Kingdom a similarly abrupt but not as great rise in the tax occurred in 1994.

The increases in taxes on oil products since the mid-1980s has prevented the price of oil products from falling in tandem with the price of crude oil; in fact, in most cases the prices of oil products have remained stable or increased. The

Figure 5.8
Indices of Crude Oil Import Costs (Thick Line) and Energy Prices for End-Users in the OECD Countries

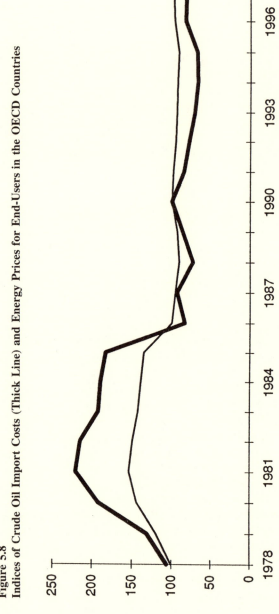

Source: International Energy Agency, *Energy Prices and Taxes* (Paris: OECD).

Figure 5.9
Taxes of Three Petroleum Products as Percent of Price

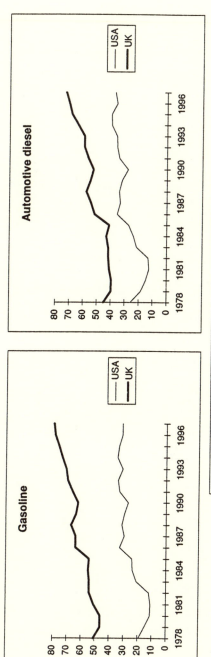

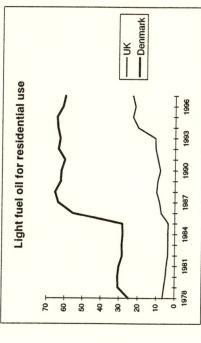

Source: International Energy Agency, *Energy Prices and Taxes* (Paris: OECD).

Figure 5.10
Indices of Prices and Taxes of Three Petroleum Products in Selected OECD Countries

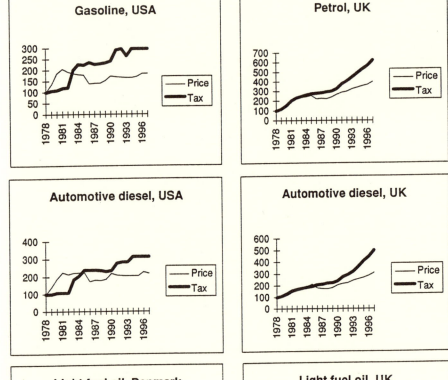

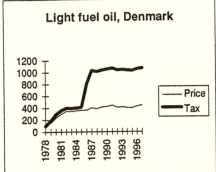
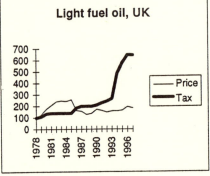

Source: International Energy Agency, *Energy Prices and Taxes* (Paris: OECD).

fall in the oil price after 1986 cannot be expected, therefore, to have increased the demand for oil dramatically. The demand for oil is determined by the final market price and a host of other factors, not by the price of crude oil. Taxes on oil products were common and high also before the first and second oil price hike in 1973 and 1980, thereby dampening the effect of higher crude oil prices on demand.

NOTES

1. See Newbery (1981) and references therein, and a comment on Newbery by Groot, Withagen, and de Zeeuew (1992), and Newbery's reply (1992).

2. An articulate exponent of this view is professor Morris Adelman of Massachusetts Institute of Technology (see Adelman, 1972, 1990, 1995).

Chapter 6

The Taxation of Oil Extraction

OIL RENTS AND OIL TAXES

As explained in the preceding chapter, production of petroleum typically generates rent, that is, profit over and above what is needed to cover all necessary costs of production. The market price of crude oil has recently been 17 to 20 US dollars per barrel while the cost of production is typically around 10 dollars a barrel in the North Sea, but much lower, and maybe as low as one dollar per barrel, in the Middle East. What makes the assessment of cost a particularly tricky business is that the oil companies pay huge sums of money for exploration for oil, most of which never results in finds that it is worthwhile to exploit. To make the exploration worthwhile the revenue from profitable finds must be high enough to compensate for all the dry holes and unprofitable finds that result from the exploration activity.

Governments in oil-producing countries usually try to obtain a substantial share of the petroleum rent through user fees or special taxes. There are both legal-philosophical and practical reasons for this. The legal-philosophical reason is that oil deposits often lie underneath public lands, so the governments in question consider themselves the rightful owners of these resources, in trust of the people who elect them. The offshore areas around the world and wildernesses such as the North Slope of Alaska belong to this category. User fees or taxes can be seen as a compensation to the landlord for allowing extraction of oil from his land, much as a tenant farmer pays a rent to an absentee landowner. Furthermore most of the oil that is produced in some countries is exported. Fees and taxes were a way for the rulers of these countries to obtain a share of the profits from the extraction, particularly before the 1970s when the production in places such as the Middle East was in the hands of foreign companies.

The practical reason for special oil extraction fees and taxes is simply that there is a lot of money to be shared; as already stated, oil rents can be very substantial. In the most profitable places very high taxes can be applied without making production economically unattractive. Indeed, there is a celebrated economic theorem to the effect that rents can be taxed at will without affecting the production activity. This is, however, often easier said than done, as the discussion to follow will make clear.

DIFFERENT FORMS OF PETROLEUM TAXES

Three major types of user fees and taxes are applied to oil extraction, fees for oil extraction licenses, royalties on production, and taxes on profits. We consider each of these in turn.

License Fees

An oil extraction license has a value to the extent it can provide profits over and above the necessary costs of extraction. This is the rental value of the site to which the license applies. The amount an oil company would be willing to pay for the license could never be higher than this. But would it be possible to entice the company to part with this sum of money to whoever allocates the license? Competitive bidding might appear to be a possibility. If many companies are interested in working the tract, competitive bidding might entice them to pay almost the whole rental value of the site to gain access to it; as long as the company retains any share of the rent it will make a net profit from extraction, and to make sure it gets the tract it will have to make a high bid, provided there is real competition among the bidders.

There are, however, problems with this. First, the extraction period is typically long and the sums of money involved are high. The company might have to borrow money at a high rate of interest to pay for the license, unless the lessor agrees to a payment schedule that stretches into the future. Second, and more importantly, the future income stream is not known but uncertain, and possibly highly uncertain. The company will be well advised to make allowance for the possibility that its expectations will turn out to be too optimistic. Therefore, each company will not be willing to pay an amount equal to the expected present value of the excess profit from the license, but something less. Seen from the lessor's perspective this method may not be a very effective way to obtain a large share of the rent of the license. There is, however, the phenomenon called the winner's curse, which may be good news for the lessor but bad news for the bidders. The bidders may have different information about the tract and may interpret the facts they have differently. The highest bidder will most likely be the one who expects the highest value of profits from the tract. That expectation may, however, be based on unrealistically optimistic beliefs about the tract. If this is true of the winner of the bidding process he will get two pieces of news,

Table 6.1
Results of Auctioning Oil Extraction Licenses on the North Slope in Alaska

Date	Acres	$/acre
December 1964	466,180	9
July 1965	403,000	15
January 1967	37,662	39
September 1969	412,548	2182

Source: G. K. Erickson, "Alaska's Petroleum Leasing Policy," *Alaska Review of Business and Economic Conditions* (July 1970).

the good news that he won and the bad news that he will go broke. The latter is, perhaps, not of any great concern for the lessor; he might be able to organize a new round of bidding for a now less promising license, but he will already have got an income based on an unduly optimistic belief about the license.

The time point when a tract is laid out for bidding will be crucial for how much the bidders are willing to bid. Before such events there usually has been some exploration, such as seismic exploration for investigating the layers underneath the surface to find out whether there are any rock formations that could contain petroleum. And if petroleum has been discovered in nearby and similar tracts, the probability that it will also be found in that particular tract will be vastly increased. The story of the auctions of oil leases on the North Slope in Alaska is telling. Table 6.1 shows the dollars per acre paid for oil leases at different time points.

Licenses for oil extraction on the continental shelf of the United States are allocated through competitive bidding but not in the North Sea. In the British part there has been some experimentation with competitive bidding but otherwise the extraction licenses have been allocated after the companies have submitted a work program. It has been alleged that this process amounts to competitive bidding in work programs, which can be shown to have disadvantageous effects (Kretzer, 1993). In the Norwegian part the licenses are allocated by the government, after the companies have indicated their preferences and submitted their plans. The Ministry of Energy sets up consortia of five to ten participating companies on each tract and appoints one as responsible for the extraction (usually referred to as the operator).

Royalties

Royalties are taxes on gross revenue and widely applied to mineral extraction around the world. In Norway royalties were introduced early on, at a rate of 8 to 16 percent for oil, depending on the size of the field, and 12.5 percent for gas. In 1986 royalties were abolished for new fields, of which more later.

Royalties would appear to be a fairly straightforward method of taxation, but since they are levied on gross revenue they may make some fields that are profitable before tax unprofitable after tax. Hence, if the royalty rate is set too high it may deprive the government of some income, as fields that the royalty makes unprofitable would not be developed. Furthermore, it will cause fields to be shut down prematurely, as the gross revenue will still be above operating costs when the revenue net of royalty has fallen to a level equal to the operating costs. For governments royalties have the attractive attribute of securing a flow of income even if the extraction is not very profitable, provided it is not shut down altogether.

Profit Taxes

Taxes that are levied on profit would appear to have the desirable property of not preventing profitable fields from being developed and operated. Such is not the case, however. The problem lies in the allowances typically made for depreciation of capital equipment. The present value of these allowances is not necessarily equal to the value initially spent on the equipment. Alternatively, the problem may be said to consist in an inadequate definition of profits; a true depreciation allowance would solve the problem, as will be demonstrated below.

To understand the nature of the problem, consider the present value (V) of a hypothetical project

$$V = -K + \sum_{t=1}^{T} \frac{R_t}{(1 + r)^t} \tag{6.1}$$

where K is the necessary investment in production equipment, T is the lifetime of the equipment, R_t is the revenue in year t net of operating costs, and r is the rate of discount applied by the company. If the project is worthwhile, the present value will be positive. Alternatively, we may calculate the value of r that makes the present value equal to zero. This is called the internal rate of return, to be compared to the discount rate applied by the company. If the internal rate of return is higher than the discount rate, the project yields a higher rate of return than required by the company and is worthwhile to undertake. Both methods give the same answer to the question of which projects are and which are not profitable.[1]

Before calculating the tax, companies are permitted to deduct depreciation allowances from their gross profit. The rules for these allowances stipulate how the initial cost incurred for the production equipment is to be spread over time. One widely used method is to deduct a certain fraction, x, every year. After $T = 1/x$ years, the entire initial outlay has been deducted. The problem is, however, that the present value of these deductions is less than the original outlay. Put another way, if each year's deduction were set aside and invested at a rate of interest equal to the rate of discount the fund accumulated after T years would not be sufficient to both provide the required return on the investment and to renew the equipment, unless T is sufficiently shorter than the lifetime of the equipment. The depreciation fund plus the market value of the not fully obsolete and worn-out equipment would in the latter case suffice to procure a new piece of equipment and to provide the required return.

To see this, consider the present value of a project with a profit tax at the rate s:

$$V = -K + \sum_{t=1}^{T}\left[R_t - s\left(R_t - \frac{K}{T}\right)\right](1 + r)^{-t} \tag{6.2}$$

Looking specifically at a project that just breaks even (a project with $V = 0$) without tax, we get

$$V = -s\left[\sum_{t=1}^{T}\frac{R_t}{(1 + r)^t} - \frac{K}{T}\sum_{t=1}^{T}\frac{1}{(1 + r)^t}\right] \tag{6.3}$$

Since $\sum_1^T (1 + r)^{-t} < T$, $V < 0$, and a project that breaks even without the tax will be unprofitable with the tax. Hence projects with a positive but low profitability will be unprofitable after tax. The profit tax therefore discourages the implementation of some profitable projects.

Note that the last term in the above expression is the present value of depreciation allowances that are invested at the going rate of interest and withdrawn after T years. The value of the first allowance will have increased to $(K/T)(1 + r)^{T-1}$ after T years, the value of the second allowance to $(K/T)(1 + r)^{T-2}$, and so forth. Hence we get the present value of all allowances as

$$\frac{K}{T}\frac{1}{(1 + r)^T}[(1 + r)^{T-1} + (1 + r)^{T-2} + \ldots + (1 + r) + 1] = \\ \frac{K}{T}\sum_{t=1}^{T}\frac{1}{(1 + r)^t} \tag{6.4}$$

As stated above, the present value of the depreciation allowances is smaller than the initial investment outlay for the capital equipment.

There are various ways of coming to terms with this problem. One is to allow investment to be written off immediately. This will work only if the company has sufficient profits to cover the entire investment cost; otherwise one must devise some form of a tax credit, to which we shall return below. In the Norwegian petroleum tax code there exists a special writeoff meant to mitigate against this problem. As we shall see, it might even go too far in the opposite direction.

THE NORWEGIAN PETROLEUM TAXES

The Development of the Tax System over Time

The Norwegian petroleum tax code has undergone major changes over the years, in response to changes in the price of oil and other factors that affect the profitability of the industry. Capturing a high share of the oil rent is a major policy objective in Norway. To achieve this, in the face of oil price gyrations and the like, the tax system itself would either have to accommodate such changes automatically, or else the system itself must be changed in response to changing circumstances. The Norwegian oil taxation policy has been successful to the extent that it has resulted in capturing something like 80 percent of the oil rent, but probably at the cost of some wasteful allocation of resources. The tax code has been marred by some built-in incentives for inefficiencies, to be discussed below, and such incentives are probably still there despite repeated revisions.

When oil prospecting began in the Norwegian sector of the North Sea the auspices were none too good. Leading geological experts in Norway thought it highly improbable that any oil would be found. One of them is reported to have offered to drink every single drop that might be discovered. It took the oil companies a long time to find anything of value; in fact, one of the very last holes that the pioneering Phillips Petroleum of Oklahoma undertook to drill turned out to be promising, after so many disappointments that the company actually tried to wriggle out of its undertaking. The Ekofisk field was found just before Christmas 1969.

Not surprisingly, the initial tax regime of the petroleum industry reflected the uncertainty surrounding the prospects on the Norwegian shelf. The government was afraid that the Norwegian shelf would not be attractive enough for foreign investors. The companies only had to pay ordinary taxes, a 10 percent royalty, and a small area fee. The "municipal" part of the ordinary company tax was reduced, to make the shelf more attractive.[2]

In 1972, when it was clear that there were profitable finds on the Norwegian shelf, the royalty rate was changed to 8 to 16 percent, depending on the maximum rate of production from a field, and 12.5 percent for gas. Then came the oil embargo in 1973, which turned the oil market upside down. In response to the ensuing leap in profitability of oil extraction a new tax code came into effect in 1975. This introduced the special petroleum tax, set at 25 percent of profits, but to sweeten the pill an extra depreciation allowance called capital uplift was introduced, amounting to 10 percent per year of the investment cost for fifteen years deductible against the tax base for the special petroleum tax. In 1979–80 came another price hike and the tax regime was made tighter; the special petroleum tax was raised to 35 percent, the capital uplift was reduced to 6.67 percent per year, and the tax credit was reduced from twelve to six months.

The price fall in 1986 made a reversal necessary. The royalty was abandoned

Table 6.2
**Income of the Norwegian Government from Various Taxes on Petroleum
Extraction in 1996, Billion Kroner**

Ordinary tax	9.9
Special tax	12.9
Royalty	6.3
Area fee, etc.	1.2
CO_2 tax	2.8
Total	33.0

Source: Statistics Norway, *Oil and Gas Activity*, 2nd Quarter, 1997.

for new fields. The capital uplift was abandoned but the companies were given a tax-free production allowance of 15 percent. The special tax was reduced to 30 percent. The final major revision came in 1992, in response to a general reduction in the company tax rate from 50 to 28 percent. To compensate for this the special tax rate was increased to 50 percent, the production allowance was abandoned, and the capital uplift was reintroduced with a rate of 5 percent for six years.

How important are these taxes? In 1996 the government obtained 33 billion kroner in tax revenue from the petroleum industry. The breakdown is shown in Table 6.2.

In addition to the tax revenue, there are the profits from the shares in the extraction licenses that the government has reserved for itself, an arrangement that started in 1985. Finally there are the dividends of Statoil, which is entirely government owned. Adding both of these to the tax revenue gives the net cash flow of petroleum revenues to the government, which in 1996 amounted to 70 billion kroner.[3] This corresponds to about 8 percent of GDP, or 16 percent of the government budget.

Incentive Problems in the Norwegian Tax System

As a simple rule of thumb, a tax system will not distort the allocation of resources if the posttax incremental income that the oil companies get from a marginal investment is equal to the share they pay of the cost of that investment. We can demonstrate this formally as follows. Let $C(q_0)$ be the cost of production capacity (q_0) and $R(q_0)$ be the present value of revenue net of operating costs for a given reservoir. C will be an increasing function of q_0 and so will R, but beyond a certain value of q_0 R will fall with q_0 because of rate sensitivity and falling reservoir pressure. The net present value of the reservoir is

$$V(q_0) = - C(q_0) + R(q_0) \tag{6.5}$$

Maximizing the present value entails

$$V'(q_0) = - C'(q_0) + R'(q_0) = 0. \tag{6.6}$$

Consider now costs and revenues with a tax on profits. The tax implies that the company keeps only a certain share of the revenue net of operating costs. Call that share s_r. Investment costs are dealt with by depreciation allowances, which decrease the tax base, so the firm pays only a certain share, s_k, of its investment costs. That share we can find by deducting the present value of the tax saved by the depreciation allowances from the investment cost and dividing by the investment cost, as will be shown below. Taking the tax system into account, the present value after tax of the reservoir will be

$$V(q_0) = - s_k C(q_0) + s_r R(q_0) \tag{6.5'}$$

and the first-order condition for maximum will be

$$V'(q_0) = - s_k C'(q_0) + s_r R'(q_0) = 0. \tag{6.6'}$$

If the tax system is not to distort the optimal solution, (6.6) and (6.6') must be satisfied for the same value of q_0. This will only happen if $s_k = s_r$, that is, if the companies retain the same share of revenue after tax as they pay of their capital investment.[4]

To illustrate, consider the current Norwegian petroleum tax system. The oil companies operating on the Norwegian continental shelf pay ordinary company tax at a rate of 28 percent, and in addition a special petroleum tax of 50 percent. The tax rate on revenue therefore is 78 percent, which gives $s_r = 0.22$. Investment costs can be written off over six years, with an equal amount per year. The tax base for the special tax is the same as for the ordinary company tax except that an extra depreciation allowance (capital uplift) of 5 percent of the investment outlay is permitted annually for six years. Assuming that the company is profitable enough to be liable to both types of taxes, we can calculate the share the company pays of each krone invested (s_k) as follows:

$$s_k = 1 - \sum_{t=1}^{6} \frac{0.78}{6}(1 + r)^{-t} - \sum_{t=1}^{6} 0.5 \cdot 0.05 (1 + r)^{-t} \tag{6.7}$$

The figure 0.78 is the sum of the rates of the ordinary tax and the special petroleum tax while 1/6-th of the krone can be written off each year for six years, so the first sum shows the present value of the tax saved by the investment through the ordinary and the special petroleum tax. In addition there is the 5 percent capital uplift. The latter sum shows the present value of the tax savings due to the capital uplift. Note that the capital uplift is good only against the special tax, the rate of which is 50 percent.

Clearly the share the company pays of its investment depends on the rate of discount. The higher the rate of discount, the more the company pays. Table

Table 6.3
The Share of Capital Costs (S_k) Paid by Companies in the Norwegian Petroleum Industry, after Taking into Account Tax Savings through Depreciation Allowances and Capital Uplift

r	With Capital Uplift	Without Capital Uplift
0.05	0.2133	0.3402
0.07	0.2612	0.3803
0.10	0.3248	0.4337
0.15	0.4134	0.5080

6.3 shows the value of s_k for selected values of r, with and without the capital uplift. As the share of revenue that the companies retain is only 0.22, we see that the capital uplift makes the tax system less distortive. The share of investment costs paid by the companies is just about equal to the share retained of the revenue at a 5 percent rate of interest, but higher at higher rates of interest. Without the capital uplift that share would be higher still. In all probability the tax system discourages investment, because the real rate of discount applied by the companies is likely to be higher than 5 percent, not least because of the uncertainty they face, which is likely to be reflected in a high discount rate.

This was not always so. The tax code in force before 1987 appears to have been quite generous, to the point of paying most of the investment costs for the companies in the form of tax savings. At that time the company tax was 50.8 percent and the special petroleum tax was 35 percent, which gives $s_r = 0.142$. The capital uplift was 6.67 percent per year over fifteen years, but no investment costs could be deducted until the field had come on stream, which greatly reduced the value of the depreciation allowance for companies that got some net profit from fields already developed. The share of capital costs paid by the firm was

$$s_k = 1 - \left[\sum_{t=1}^{6} \frac{0.858}{6}(1 + r)^{-t} + \sum_{t=1}^{15} \frac{0.35}{15}(1 + r)^{-t} \right](1 + r)^{-h} \qquad (6.7')$$

The construction of this formula is analogous to the one above except for the last term, which accounts for the time delay (h) between the investment expenditure and the startup of production from the field. Table 6.4 shows some values of s_k for selected values of r and h.

In nominal terms the system was overly generous; for each krone invested the tax saved was no less than 1.2 kroner. Since there often is a long time delay between investment and the start of production and the oil companies probably apply a discount rate of 10 to 20 percent, the present value of the tax savings was low enough to make s_k positive, but not much higher than 0.05 to 0.1, so the oil companies were not paying more than 10 percent or so of their investment costs. For investments undertaken late in the development phase they hardly paid anything at all; the government paid most of it in the form of lost

Table 6.4
The Share of Capital Costs (s_k) Paid by Companies in the Norwegian Petroleum Industry according to the Pre-1987 Tax Code, after Taking into Account Tax Savings through Depreciation Allowances and Capital Uplift

r	$h = 0$	$h = 5$
0.0	−.2	−.2
0.1	−.01	0.05
0.2	0.07	0.11

tax revenue. The incentives for the companies to keep down costs were correspondingly weak. Not surprisingly, these were years with high costs and cost overruns in the North Sea, so much so that a special committee was appointed by the government to study the problem. This illustrates a general problem associated with taxing profits at high rates; the higher the tax rate, the more of the cost of production and investment is paid by the government in the form of lost tax revenue, and the incentives to keep down costs are correspondingly weakened.

The difference between the ordinary tax rate (28 percent) and the rate applied to the petroleum industry (78 percent), together with the fact that interest payments on loans are deductible costs, provides incentives for diversified companies to attribute as much as possible of their interest costs to their petroleum activities. This will tend to lower the tax revenue. Furthermore it will distort the incentives to invest; an investment that is not profitable before tax may very well turn out to be so after tax, if the interest cost can be attributed to the petroleum branch of the company. Suppose the company borrows ten million kroner at a 10 percent rate of interest and invests it in shares that yield only 8 percent. With the ordinary company tax the company would effectively pay 720,000 kroner as interest cost and keep 576,000 of the return on the shares, which is not profitable. But if the interest cost could be attributed to the petroleum sector the company would only pay 220,000 as interest cost, because of the 50 percent special petroleum tax, and would thus come out ahead. Hence such companies might be enticed to invest in projects that are not profitable at the going rate of interest, provided they can unload the financial cost on the petroleum-based part of their activities.

THE RESOURCE RENT TAX

Finding a tax that is neutral (in the absence of other distortions) with respect to profitability is simple enough in principle but such taxes may be difficult to put into practice. A cash flow tax is a neutral tax, as already mentioned; the companies can deduct their investment costs immediately from their tax base, which automatically makes the present value of their depreciation allowance equal to the actual investment costs incurred. To make this work the company would have to make a profit that exceeds the investment cost, which for expensive and long-term projects like oil extraction is none too likely.

Alternatively the company would have to receive payments instead of paying taxes whenever the cash flow is negative, which governments do not like, or get a tax credit that can be carried forward with interest. This raises the question; what would be the appropriate rate of interest to use? In order not to distort the company's decision it would have to be the rate used by the company itself, but information on such rates is not immediately available and is even a well-guarded secret by some companies. The company rate may, however, not be the one that would be desirable from a social point of view, which raises a further question how that might be compensated for, a point we shall not dwell on here.

The so-called resource rent tax, proposed by Garnaut and Ross (1975), is an attempt at constructing a neutral and practicable tax system. It is best explained by an example similar to the one originally used by Garnaut and Ross and set out in Table 6.5. The investment project in Table 6.5 stretches over twenty years, with the first three being the investment phase, followed by seventeen years of steady income. The rate of interest is set at 10 percent. Both cost and income are treated as if they accrued at the end of each year. The investment cost is carried forward at a 10 percent rate of interest, assumed to be the discount rate applied by the firm, so at the end of year 2 the investment cost of the first year has grown to 110, to which another 100 is added, and so forth. At the end of year 4 the accumulated cost has grown to 364 but from this we subtract the first revenue and end up with 314 as accumulated loss, and so forth. This continues until all the initial investment costs have been paid for, with the 10 percent stipulated rate of return on invested capital. But even if the project is profitable—it has an internal rate of return just below 13 percent—it takes a long time for any tax revenue to emerge; this does not happen until year 15 when the investors have got back all their capital with the necessary return, which might be a long time for a minister of finance to wait for something to show.

There are alternative ways of coping with the problem of ensuring that the investors get their money back with the necessary return while making it possible to gather some tax income earlier. The essence of the solution is to make the present value of the depreciation allowances equal to the original investment. This will enable investors to recoup the money they invested with the required rate of return (equal to the rate of discount used in calculating the present value of the depreciation allowances). One possible method is to allow a constant value to be deducted from profits every year in such a way that the present value of deductions is equal to the initial investment. The annual amount to be deducted can be calculated to[5]

$$rK\left[1 + \frac{1}{(1 + r)^T - 1}\right]$$

Table 6.5
The Resource Rent Tax: An Example

Year	Investment	Net revenue	Accumulated loss	Profit	Tax (50%)
1	100		100		
2	100		210		
3	100		331		
4		50	314		
5		50	296		
6		50	275		
7		50	253		
8		50	228		
9		50	201		
10		50	171		
11		50	138		
12		50	102		
13		50	62		
14		50	18		
15		50		30	15
16		50		50	25
17		50		50	25
18		50		50	25
19		50		50	25
20		50		50	25

where K is the amount invested. At the end of year 3, this cost is 331.1 at a 10 percent rate of interest. This gives a depreciation allowance of 41.28 for year 4 through 20. It is readily checked that the present value of this over seventeen years (discounted to the end of year 3) is precisely 331.1. The taxable income each year will be 8.72, which gives a tax income of 4.36 from year 4 onward with a tax rate of 0.5. This could be a better tax profile for a minister of finance than the previous one, but it is readily checked that the present value of both is the same (34.9) when discounted to the end of year 3.

The advantage of the resource rent tax is that it makes it possible to cope with high variability of oil prices and revenues without changing the tax code. This is so because the resource rent tax is a tax on pure profit whereas ordinary "profit" taxes are not, because of the way investment costs are treated. Costly and cumbersome changes in the tax code, as have taken place in Norway and to an even greater extent in Britain, could be avoided. The resource rent tax also offers the companies a greater certainty with respect to tax regimes. On the other hand, the uncertainty of the present tax code is perhaps more apparent than real; companies know from experience that governments will not sit by idly if their present tax system does not give them an appropriate share in windfall profits such as emerge from unforeseen price rises. Conversely they are forced to revise their rules in the opposite direction if prices fall unexpectedly, in order to ensure that their countries remain attractive for investors. The above-mentioned changes in the Norwegian tax code are an example of such revisions. The companies may very well have anticipated those changes.

But the resource rent tax also has problems of its own. The discount rate applied by the companies is not known. If guessed wrongly, the resource rent tax will either be too generous and encourage investment, or the opposite. Furthermore, it depends on correct information about revenues and costs. Many avenues are open for international oil companies to manipulate these figures. By an appropriate pricing of goods and services between their different divisions and subsidiaries they can realize their profit in countries with light tax regimes and make sure they earn little profit in places with cash hungry governments. In fact this is not a problem only for the resource rent tax; all profit taxes encounter this problem.

NOTES

1. It is possible to get more than one solution for the internal rate of return, in which case the method becomes a bit ambiguous, but this happens only if the net revenue stream changes sign more than once.

2. There is no municipal tax on oil extraction in Norway; the "municipal" part accrues to the central government like all other taxes on petroleum.

3. Source: The National Budget 1998 (Stortingsmelding No. 1, 1997–98). The government also holds a large number of shares in Norsk Hydro, but this company is engaged in many other activities besides oil and gas production and its dividends are not included in the net cash flow of government petroleum revenues.

4. The equality $s_r = s_k$ will ensure optimality only if there are no other distortions causing deviations from optimality. If, for example, the companies apply a higher discount rate than desirable from a social point of view they will invest less in field development than desirable. This can be corrected by making s_k lower than s_r, which stimulates investment.

5. This formula can be found as follows. The total capital cost each year is the interest on investment costs not yet deducted plus the deduction that year. Hence the following capital costs:

year 1: $rK + A_1$,

year 2: $r(K - A_1) + A_2$

. . .

year t: $r(K - \sum_1^{t-1} A_i) + A_t$.

Equality of capital costs across all years implies

$$r(K - \sum_{i=1}^{t-2} A_i) + A_{t-1} = r(K - \sum_{i=1}^{t-1} A_i) + A_t$$

or $A_{t-1} = -rA_{t-1} + A_t$, so that $A_t = A_{t-1}(1 + r)$. Now, $\sum A_t = K$, so that $A_1 + A_1(1 + r) + A_1(1 + r)^2 + \ldots + A_1(1 + r)^{T-1} = K$, which implies $A_1 = rK/((1 + r)^T - 1)$. Since all annual costs are equal, the annual cost is $rK[1 + 1/((1 + r)^T - 1)]$.

Chapter 7

The Management of Petroleum Wealth

PETROLEUM FUNDS

Oil and gas, like other mineral resources, are nonrenewable. It may very well be, as argued elsewhere, that the price formation for oil and gas takes little notice of this fact; it may well be that for all practical purposes oil and gas resources are so plentiful that we may regard them as infinite, with the long-term supply price being determined by what it takes to maintain a reasonable inventory of these resources. However that may be, things may look quite differently when seen from the point of view of a particular oil-producing country, state, or province. The reserves in a particular country or region may become depleted over the lifetime of a generation or two even if reserves may be discovered elsewhere so that the global reserves can be maintained.

A country that depletes its reserves over a time span of a generation or so faces the choice of letting that generation use up the oil wealth for its own benefit or giving future generations a share in these resources. But how can future generations benefit from a nonrenewable resource that will have been depleted before they were even born? Clearly this is possible only by transforming the nonrenewable resource into a renewable one. The way to do this is to invest the rents earned from extracting oil and gas in ways that increase the production capacity at home or abroad and so raise the standards of living for the present as well as future generations.

A practical device to accomplish this transformation is an investment fund into which the petroleum rent, or an appropriate share thereof, is channeled. The fund would invest in financial assets—company stock and bonds, government bonds and real estate—behind which is productive real capital in the form of machines, buildings, infrastructure, or knowledge embodied in people. A hypothetical example of such a fund is shown in Table 7.1. The extraction of petroleum is assumed to last for fifteen years. The second column shows the petroleum revenue net of all costs. The revenue is assumed to be small initially, with a peak in the middle, and tapering off toward the end of the extraction period, like the production profile of an oil field discussed in Chapter 4.

Suppose it is possible, in this hypothetical economy, to achieve a real rate of return of 7 percent on invested capital.[1] Discounted at this rate, the present value of the petroleum revenue is 408.78, as shown in the table. This is the petroleum wealth. If this amount were invested it would be possible to earn a return of 7 percent, which amounts to 28.61 per year. This is the part of the petroleum revenue we could use each year and still keep the petroleum wealth intact. By not using more than this the petroleum wealth could benefit not only the present but also all future generations.

Table 7.1 also shows the investment fund we need to build up if we follow the strategy of using 7 percent of the petroleum wealth each year. To begin with, the fund accumulates a substantial debt. The rents in the first year are only 1 while 28.61 are being used, so 27.61 need to be borrowed and the net balance of the fund at the end of the first year is −27.61. In the second year 28.61 are again used but there is now an income of 10, so only 18.61 are borrowed this time. At the end of year 2 the accumulated debt is 48.16, which includes an interest charge of 1.93 (7 percent) on the debt carried over from year 1. But in year 5 the debt stops accumulating; the petroleum revenue has by then outgrown the amount being used every year. From year 7 to 15 a fund is built up, and at the end of year 15 when the oil extraction is finished the fund amounts to 408.78, which is equal to the present value of the oil revenues. From then on the 7 percent annual return on the fund can be used without depleting the fund, a policy that can go on forever if the rate of return is maintained. The nonrenewable petroleum resource has been transformed into a renewable one by the device of an investment fund.

As Table 7.1 shows, the use of the "oil money" initially exceeds the oil revenue. This is not due, however, to the present generation (if we can talk about different generations in a twenty-year perspective) taking advantage of future generations and using the petroleum wealth for itself but to a smoothing of the consumption profile. The petroleum wealth, presumably discovered immediately prior to year 1, makes the country permanently richer if it is used appropriately, as in this example. If the use of the oil wealth is not smoothed in this way and we set ourselves the goal of building up a fund of 408.78 before using any of the oil wealth, the consumption would be smaller initially and would reach a peak in the middle of the extraction period when the oil money is abundant. This would be to the disadvantage of those who are old in the beginning of the extraction period and to the benefit of those who happen to be alive in the peak years.

Needless to say, many practical problems are glossed over by this simple example. Oil prices are, as we have seen, volatile and can double or triple, or

Table 7.1
A Hypothetical Petroleum Fund

Year	Revenue	Discount factor (r = 0.07)	Present value of revenue	Deposit to fund	Fund balance beginning of year	Fund yield	Fund balance end of year
1	1	0.935	0.935	-27.615	0.000	0.000	-27.615
2	10	0.873	8.734	-18.615	-27.615	-1.933	-48.162
3	20	0.816	16.326	-8.615	-48.162	-3.371	-60.148
4	30	0.763	22.887	1.385	-60.148	-4.210	-62.973
5	50	0.713	35.649	21.385	-62.973	-4.408	-45.996
6	70	0.666	46.644	41.385	-45.996	-3.220	-7.830
7	100	0.623	62.275	71.385	-7.830	-0.548	63.007
8	100	0.582	58.201	71.385	63.007	4.410	138.803
9	100	0.544	54.393	71.385	138.803	9.716	219.904
10	80	0.508	40.668	51.385	219.904	15.393	286.683
11	60	0.475	28.506	31.385	286.683	20.068	338.136
12	40	0.444	17.760	11.385	338.136	23.670	373.191
13	20	0.415	8.299	-8.615	373.191	26.123	390.700
14	10	0.388	3.878	-18.615	390.700	27.349	399.434
15	10	0.362	3.624	-18.615	399.434	27.960	408.780
16	0	0.339	0.000	-28.615	408.780	28.615	408.780
17	0	0.317	0.000	-28.615	408.780	28.615	408.780
18	0	0.296	0.000	-28.615	408.780	28.615	408.780
19	0	0.277	0.000	-28.615	408.780	28.615	408.780
20	0	0.258	0.000	-28.615	408.780	28.615	408.780
Present value of revenue (V)			408.780				
Annual yield of V at 7 per cent			28.615				

be reduced by a half, quite abruptly. This will substantially affect the petroleum wealth and change the amount that should be set aside accordingly. How should this risk be dealt with? The downside risk of low revenues is probably more difficult to live with than the upside risk of high revenues. If so, the accumulation of the oil fund should be greater than otherwise, so as to be better able to face the low incomes resulting from a falling price of oil. We return to the question of risky oil prices, but in a different setting, in the last part of this chapter.

AN EXAMPLE: THE ALASKA PERMANENT FUND

How have nations responded to the challenge of managing their petroleum wealth? Have they built up investment funds or have they used up all the rents immediately for the benefit of one or a few generations? Some countries at least have set aside a substantial amount of their oil rents for future use. The government of Kuwait was able to keep itself going in exile for several months of Iraqi occupation in 1990–91 by the oil revenue it had invested abroad. Even if the fund was severely depleted during the occupation, roughly 20 percent of the Kuwaiti national income in 1993 was derived from assets abroad.[2] The Canadian province of Alberta and the American state of Alaska have both set up investment funds into which a substantial share of the oil revenue has been channeled. Of these the Alaska Permanent Fund appears to have been particularly successful.

The Alaska Permanent Fund was set up explicitly to make sure that the revenues the state of Alaska gets from the oil extraction on the North Slope benefited all Alaskans. The fund is a device for spreading the oil revenue among the inhabitants of the state, over time and among individuals at any given time. Spreading the benefits over time is accomplished by building up a fund by depositing at least 25 percent of the oil revenue the state gets into the fund, using only the real return on the fund (the return after setting aside what is necessary to compensate for inflation). Spreading the benefits among the inhabitants of the state is accomplished by distributing the disposable return on the fund by sending a check to each person living in the state. In recent years this amount has been about one thousand dollars per year for every man, woman, and child in the state. In the fiscal year of 1996 36 percent of the fund's income was distributed as dividends, 23 percent was needed for inflation proofing, and the remaining 41 percent was added to the principal through a special appropriation.[3]

The way a temporary income stream from a nonrenewable resource can be transformed into a permanent income stream by an investment fund is well illustrated by the profiles of the income of the Permanent Fund and the oil revenues of the state of Alaska shown in Figure 7.1. The oil revenues of the state are double peaked and are expected to taper off gradually from the mid-1990s onward. The Permanent Fund has been built up gradually (see Figure

Figure 7.1
Oil Revenues of the State of Alaska and the Income of the Alaska Permanent Fund

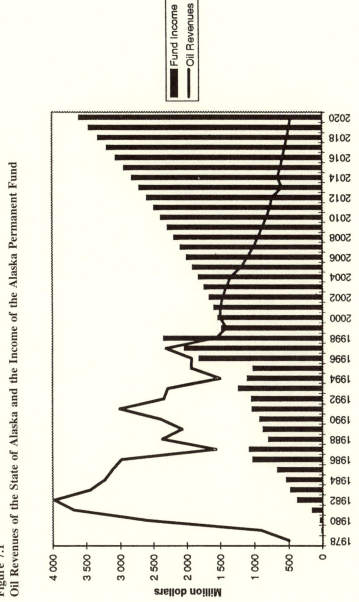

Source: Alaska Permanent Fund.

7.2) and its income has increased, albeit with interruptions, and is expected to increase further, reaching almost the same value in 2010 as the peak of state oil revenues in 1981.[4]

From an institutional perspective the Permanent Fund is a very interesting and possibly unique construction. First, great care was taken to make the fund permanent and independent of short-sighted political manipulation. Twenty-five percent of the oil revenues of the state were earmarked for the fund, and none of its income can be disposed of until the real value of the fund has been secured by allocating the necessary share of the income to the principal to compensate for inflation. The principal itself cannot be touched. The statutes of the fund are enshrined in the constitution of Alaska, which can only be changed by a popular vote. Hence the fund is independent of the legislature and its possible temptation to use the fund to solve the fiscal problems of the state.

Second, incentive mechanisms were devised to make the Alaskan public interested in preserving the fund. As stated above, a part of the income of the fund, after allowance for inflation has been made, is distributed to all people living in the state. For low-income families in particular, a thousand dollars per head makes a difference. Proposals to erode the independence of the fund and make it subject to the legislature are unlikely to succeed, as people might fear that the cash payment would disappear and be replaced by uncertain and intangible benefits arising from the spending decisions of the legislature.

In terms of financial management the fund appears to have been reasonably successful. The real rate of return on the fund has been well above 5 percent on the average since its inception. The fund is committed to a rather conservative investment strategy by its statutes, making safety of principal a priority. In 1996 almost a half (43 percent) of the fund's assets were fixed-income bonds (mainly government and corporate bonds), followed by U.S. stocks (38 percent), international stocks (11 percent), and real estate (8 percent). The fund is run by a small and effective administration; its operating expenses in the fiscal year ending in 1996 were 28 million dollars, 1.5 percent of total revenue.

The Permanent Fund is in marked contrast to the Norwegian Petroleum Fund. The latter is in no way an independent institution but only an account in the Bank of Norway, which has been entrusted with the task of managing the fund; the fund has neither an independent administration nor a board of governors. The purpose of this construction is making the use of the government's oil revenue more explicit; the revenue is initially deposited in the fund, and the Norwegian parliament must make an explicit decision about withdrawing money from the fund. Furthermore, if the government budget is passed with a deficit it will automatically be covered from the fund's assets, as long as there are any. Whether this will restrain legislators and promote greater care in public finances remains to be seen. Their resolve has not yet been put to test; although the fund was established by law in 1990 its balance remained zero until 1996 when deposits for the first time exceeded withdrawals. The fund has increased extremely rapidly since then; it was 46.3 billion kroner at the end of 1996 and

Figure 7.2
Growth of the Alaska Permanent Fund, 1976–97

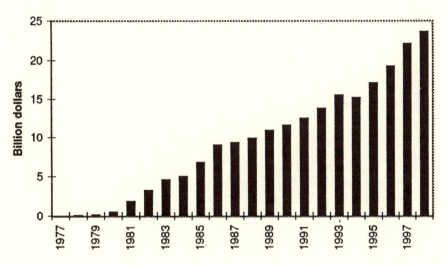

Source: Alaska Permanent Fund.

had grown to 113 billion at the end of 1997. Not only did the net cash flow of the government from the petroleum industry increase extremely rapidly in these years, they were also boom years for the Norwegian economy. If, in the economic storms that are bound to come sooner or later, Norwegian parliamentarians choose to act like a spendthrift who keeps a second account in the bank and makes the explicit decision to withdraw money from it when the ordinary account has been depleted, the fund will have served little purpose.

HOW MUCH OF THE OIL RENTS SHOULD BE SET ASIDE?

What proportion of the oil rent should be set aside and deposited in an investment fund? This depends on a number of circumstances: the level of affluence of the country at the time the petroleum rents start to flow, the expected population growth and economic growth from other sources, and so forth. These are important but difficult questions to analyze in any detail. To put them in perspective, let us look at a simple model of economic growth, with and without petroleum resources. Let $u(c)$ be the rate of utility derived from the current flow of consumption, c. Let utility be discounted at the rate r, which denotes social time preference, and let the goal of economic policy be maximization of the discounted sum of utilities from now to eternity. In continuous time the discounted sum of utilities is

$$\int_0^\infty u(c_t)e^{-rt}dt,\tag{7.1}$$

Consumption is the difference between what is produced and what is set aside for increasing and maintaining the stock of production capital, k. If capital decreases at a constant rate a, the growth rate of the capital stock will be

$$dk/dt = f(k_t) - c_t - ak_t,\tag{7.2}$$

where $f(k)$ is a function relating the flow of production to the stock of capital. We shall ignore the role of labor and implicitly assume that the supply of labor is constant and that there is no technological progress, to focus on the issue of use of petroleum wealth.

To find the appropriate path of consumption, we can use the maximum principle.[5] This involves maximizing the so-called Hamiltonian function at each point in time, with respect to c. The Hamiltonian function consists of the integrand in the objective functional (7.1), plus the function describing the development of the stock variable(s) (7.2) multiplied by an auxiliary variable λ that may change over time:

$$H_t = u(c_t)e^{-rt} + \lambda_t[f(k_t) - c_t - ak_t].\tag{7.3}$$

Henceforth the time subscripts will be dropped and subscripts used to denote partial derivatives.

Maximization of the Hamiltonian implies

$$H_c = u'(c)e^{-rt} - \lambda = 0,\tag{7.4}$$

where the prime denotes the first derivative.

A second necessary condition associated with the maximum principle says that the rate of change of the auxiliary variable (λ) will be equal to the partial derivative of the Hamiltonian function with respect to the associated stock variable:

$$d\lambda/dt = - H_k = -\lambda[f'(k) - a],\tag{7.5}$$

From (7.4) we get

$$d\lambda/dt = [-ru'(c) + u''(c)(dc/dt)]e^{-rt}\tag{7.6}$$

Equations (7.5) and (7.6) imply

Figure 7.3
Optimum Capital Stock under the Golden Rule and the Modified Golden Rule

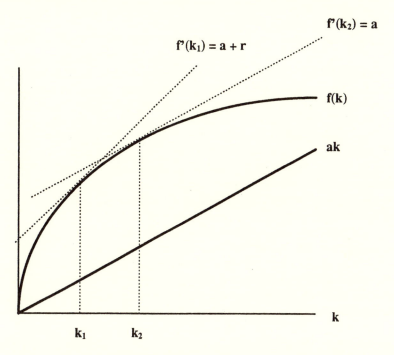

Note: The golden rule (do as you would be done by) prescribes building up the capital stock to the level k_2 where the consumption per capita, $f(k) - ak$, is maximized. The modified golden rule says that the future does not count quite as much as the present (time preference at the rate r), so we need not build up the capital stock any further that to k_1 (or we can draw it down to k_1), in which case a greater consumption now and in the near future is possible than under the golden rule.

$$f(k) = r + a - (u''/u')(dc/dt). \tag{7.7}$$

This is a well-known result from the theory of economic growth. Note in particular that if the economy approaches a steady state where no variable is changing, the optimum capital stock will be given by $f'(k) = a + r$; that is, the marginal productivity of capital should be equal to the rate of depreciation plus the rate of time preference. This is called the modified golden rule; the golden rule involves maximizing consumption per capita, or total consumption if we assume a stationary population. This is illustrated in Figure 7.3. The line ak shows the replacement of worn-out capital that is necessary to maintain the capital stock intact. The difference between this line and the production function $f(k)$ represents consumption, which is maximized when the slope of the function $f(k)$ is equal to the slope of the line ak. This is the golden rule; each generation leaves behind a capital stock (k_2) that makes it possible for the next generation to achieve the same consumption as the outgoing generation did; in the parlance that currently is fashionable this might be called the sustainability rule instead of the biblically inspired golden rule.

The modified golden rule involves a steady state with a lesser consumption per capita (capital stock k_1) than the golden rule, due to the time preference rate r. Under the modified golden rule the current generation, or generations close to the present, exploit future generations by increasing their consumption instead of investing in capital equipment that would maximize the consumption of future generations. The legitimacy of a positive social rate of time preference has been called into question by many; societies prevail while individuals come and go. For an individual a positive rate of time preference makes perfect sense: people are impatient by nature and each person faces a certain mortality risk. But societies are not immortal either; there is, for example, the slight risk that the earth might be hit by a deadly meteorite or that a nuclear war might put an end to civilization as we know it, by accident if not deliberately.

But what happens before we get to the steady state? That depends on whether the capital stock we are endowed with is greater or smaller than the steady-state stock. If it is greater we have overinvested in the past, and we can increase our consumption temporarily while we adjust the capital stock downward. A more interesting and relevant case is where we start with less than the optimal steady-state capital stock. Then we have to restrain our present consumption in order to build up the capital stock, which should be done gradually; there are limits to how much we should decrease current consumption in order to consume more later.

As an illustration, we can look at a simple numerical example. We shall use the following specifications of the utility function and the production function:

$$u(c) = lnc \qquad\qquad\qquad\qquad (7.8)$$

$$f(k) = 2Ak^{0.5} \qquad\qquad\qquad\qquad (7.9)$$

where A is a constant, calibrated so that the optimum steady-state capital stock is equal to 1. Figure 7.4 shows the optimal development of consumption and the capital stock when we start with a capital stock of 0.5, one-half of the optimal steady-state stock. Essentially what we need to determine is the initial flow of consumption. Once this is done, the differential equation for consumption implicitly given by (7.7) will trace out the path of future consumption. The solution we get is approximately $c_0 = 0.12$. As seen from Figure 7.4, the optimum paths of the capital stock and consumption involve a gradual approach toward the steady-state values $k = 1$ and $c = 0.2$.

But what difference does a discovery of petroleum or other mineral wealth make? Let s denote a mineral resource endowment that is suddenly discovered at a certain point in time. Suppose, despite the Hotelling rule, that the price of

Figure 7.4
Development of Capital and Consumption without Oil

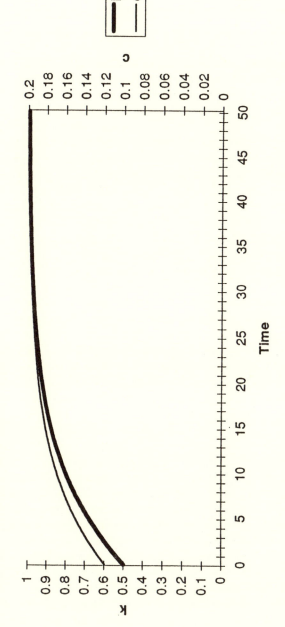

Note: Optimal growth of capital (k) and consumption (c) toward an optimal steady state of $k^{\circ} = 1$, $c^{\circ} = 0.2$, with a utility function $u(c) = lnc$, a constant discount rate of utility (0.05), a production function $y = 0.3k^{0.5}$, and a constant rate of depreciation of capital (0.1). With a given initial value of capital $k_0 = 0.5$ the optimal c_0 is approximately 0.12, with the future development of c being given by the differential equation $dc/dt = c[0.15k^{-0.5} - 0.05 - 0.1]$ and of k by $dk/dt = 0.3k^{0.5} - c - 0.1k$.

the extracted mineral relative to consumption is constant over time. Let the rate of extraction of the mineral be q. Suppose that the mineral wealth can costlessly be transformed into consumption or production capital. Equation (7.2) now becomes

$$dk/dt = f(k_t) + q_t - c_t - ak_t. \tag{7.2'}$$

The mineral resource introduces a new stock variable s with a rate of change $ds/dt = -q$, so we get an additional term in the Hamiltonian function:

$$H_t = u(c_t)e^{-rt} + \lambda_t[f(k_t) + q_t - c_t - ak_t] - \gamma_t q_t. \tag{7.3'}$$

Equation (7.4) will not be affected, which means that the optimum steady-state capital stock is not affected at all by the new wealth. What will be affected are the paths of consumption and capital accumulation by which the steady state is approached. For q we get a new set of equations telling us what the optimum extraction profile will be like. First, at each point in time, q should be chosen so as to maximize the Hamiltonian function. The partial derivative of the Hamiltonian function with respect to q is

$$H_q = \lambda_t - \gamma_t. \tag{7.10}$$

From this we see that H_q is given and independent of q at any point in time, which leaves us with three possibilities. First, H_q might be equal to zero, in which case the optimum value of q is indeterminate. Otherwise, either $H_q > 0$ and q should be made as large as possible, or $H_q < 0$ and q should be made as small as possible, which is zero. But how does H_q change over time? In the optimum steady state, $f'(k) = a + r$ (see Figure 7.3 and Equation [7.7]), so as long as we have not reached that state we know from Equation (7.5) that λ must be falling. The rate of change in γ is given by

$$d\gamma/dt = -H_s = 0, \tag{7.11}$$

so H_q must be falling over time. The only possible solution is that $H_q > 0$ at time zero and that all the mineral wealth should be extracted immediately at that time.

Returning to our numerical example, Figure 7.5 shows the optimum development of the capital stock and consumption when at time zero a mineral wealth of 0.1 is discovered (10 percent of the optimum steady-state capital stock or one-fifth of the initial capital stock). All the mineral is extracted immediately but the initial consumption does not increase by nearly that much; the value of c_0 is increased from approximately 0.12 to approximately 0.136 while most of the mineral wealth (0.084) is used for augmenting the capital stock. Hence the discovery of the mineral wealth eases the burden of capital accumulation and

Figure 7.5
Development of Capital and Consumption with Oil

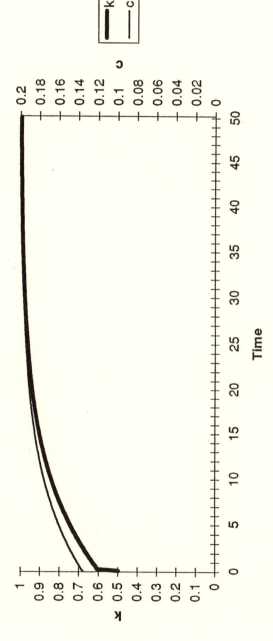

Note: Optimal growth of capital (k) and consumption (c) toward an optimal steady state of $k^\circ = 1$, $c^\circ = 0.2$ when a nonrenewable resource of 0.1 is discovered and extracted at time zero. The optimum initial consumption is raised by 0.016, while the remaining 0.084 is used to increase the capital stock.

allows some increase in the current consumption, but only a fraction of the mineral wealth is used for this latter purpose, the larger part being transformed into production capital to increase future consumption. How much of the mineral wealth should be used for immediate consumption depends on how close we are to the steady state; if we have achieved it already all of the mineral wealth should be used for consumption, since more production capital is of no use anyway.

Needless to say, this simple approach to the problem of optimally using mineral wealth glosses over many important aspects. Extraction of minerals is not a costless activity; as discussed elsewhere, it requires the construction and operation of expensive and elaborate structures. For that reason extraction would be spread over time, the more so the more expensive it is. If the relative price of mineral wealth increases over time, as the Hotelling rule would lead us to believe, the timing of the extraction depends on what gives the highest rate of return, value appreciation in the ground or investment in financial or production capital. Furthermore, mineral resources play a vital role in the production of goods and services and are seldom used directly. This raises the question of how nonrenewable mineral resources can over time be replaced by renewable resources and production capital in the production process.

UNCERTAIN PRICES AND THE OPTIMUM RATE OF EXTRACTION

How will uncertainty with respect to future oil prices affect the optimal rate of extraction? Simple intuition provides arguments that pull both ways. One bird in the hand is better than two in the bush, so we might as well grab our wealth while we still have it for sure. This would speak for an accelerated extraction in response to increased uncertainty about oil prices. But there are other arguments. Perhaps we are crucially dependent on the future oil wealth. If the future prices fall we would need a greater quantity of oil in the future to pay for whatever we need. This argument pulls the other way and speaks for less extraction at the present in response to increased uncertainty about the future.

The empirical relevance of this latter argument may not seem very great, as there are ways to transform an uncertain oil wealth into a more secure one by investing in real or financial assets. This, however, implies that governments resist the temptation of spending all the oil wealth on increased consumption as it is extracted instead of investing it to increase consumption later. Those who doubt the ability of governments to manage petroleum wealth prudently sometimes argue that whatever we don't need for our immediate consumption had better be kept under the ground and out of reach of politicians.

The ambiguity of the two arguments presented above is reflected in the fact that utility functions that are plausible representations of underlying preferences give different answers to the question how uncertainty affects the rate of extraction, given that the only way to save oil wealth is to keep it in the ground.

We shall deal with the problem in the following way. Assume a given initial amount of petroleum wealth, denoted by w. This wealth can be consumed over two periods. What is left over from period 1 will be consumed in period 2, so that

$$c_1 = c, \; c_2 = w - c, \tag{7.12}$$

where c_i is the consumption in period i. The utility function is identical in both periods but the second-period utility is discounted at the rate r, so the optimum use of the wealth is given by maximizing the present value of utilities over the two periods:[6]

$$\text{Max } V = u_1(c) + u_2(w - c)/(1 + r). \tag{7.13}$$

The first-order condition is

$$u_1'(c) = u_2'(w - c)/(1 + r). \tag{7.14}$$

Uncertainty with respect to future oil prices implies that the real value of what is left over of the oil wealth in the second period is uncertain. We shall deal with this by introducing an oil price variable, x, the value of which is uncertain. Assuming that the utility function adequately reflects the attitude to risk, we can proceed by maximizing the expected present value of utilities:[7]

$$\text{Max } EV = u_1(c) + (1 + r)^{-1}Eu_2[(w - c)x] \tag{7.13'}$$

where E denotes expected value. The first-order condition becomes, similarly

$$u_1'(c) = (1 + r)^{-1}Exu_2'[(w - c)x]. \tag{7.14'}$$

Now compare two cases, with and without uncertainty, such that in the absence of uncertainty $x = 1$ while with uncertainty $Ex = 1$. Thus the uncertainty case is equivalent to the certainty case in the sense that the expected oil wealth in the case of price uncertainty is equal to the wealth in the certainty case. How will uncertainty affect c, the rate of extraction in the first period? Denote the rates with and without uncertainty as c_d for the rate in the deterministic case and c_u for the rate in the uncertain case. Is $c_d > c_u$, or is it the other way around?

Clearly, the answer depends on which is greater, for a given c, $Exu_2'[(w - c)x]$ or $u_2'(w - c)$. To see what this means, note that the random variation in u_2' is due to the variability of x once c has been determined. We can therefore multiply both terms by the constant $(w - c)$ and the inequality will be preserved. Furthermore, $Ex = 1$ by assumption. Now define $f(s) = su'(s)$ where $s = (w - c)x$. If $f(s)$ is a concave function, it follows from Jensen's inequality that $Ef(s) < f(Es)$, or

$$Exu_2'[(w - c)x] < u_2'(w - c). \tag{7.15}$$

If this inequality holds, it means that $w - c$ must become smaller as a result of uncertainty about x. Due to the concavity of the utility function ($u'(c) > 0$ and $u"(c) < 0$) the right-hand side of (7.14') would rise, and the left-hand side would become smaller, so by finding a suitable $c_u > c_d$ the equality in (7.14') would be established. Hence, uncertain oil prices would raise the initial rate of extraction if the function $f(s) = su'(s)$ is concave.

What does this mean? From the definition of $f(s)$ we have

$$f(s) = su'(s), f'(s) = u'(s) + su"(s), f"(s) = 2u"(s) + su"'(s). \tag{7.16}$$

The function $f(s)$ will be concave if $f"(s) < 0$. Since $u"(s) < 0$ by assumption, it is necessary that $u"'(s) > 0$ in order that $f(s)$ be convex, in which case uncertainty about future oil prices would lead to a lower initial rate of extraction.

We shall illustrate with three popular utility functions, a quadratic utility function, a logarithmic utility function, and one where the elasticity of marginal utility (z) is constant. Using the subscripts q, l, and ce to denote the type of function we get

$$u_q = 1 - 0.5(c^*-c)^2 \quad u_1 = lnc \quad u_{ce} = -c^{1-z} \tag{7.17}$$

with the derivatives

$$u_q' = c^* - c \qquad u_l' = 1/c \qquad u_{ce}' = -(1 - z)c^{-z} \tag{7.18}$$

$$u_q" = -1 \qquad u_l" = -1/c^2 \qquad u_{ce}" = z(1 - z)c^{-1-z} \tag{7.19}$$

$$u_q"' = 0 \qquad u_l"' = 2/c^3 \qquad u_{ce}"' = -z(1 - z)(z + 1)c^{-2-z} \tag{7.20}$$

so, for the function $f(s)$ we get

$$f_q"(s) = -2 \qquad f_1"(s) = 0 \qquad f_{ce}"(s) = z(1 - z)^2 s^{-1-z} > 0. \tag{7.21}$$

Hence, from three popular and plausible utility functions we get three different results with respect to how uncertainty about future oil prices would affect the initial rate of extraction, one in which it would rise (the quadratic function), one in which it would not be affected at all (the logarithmic function), and one in which it would fall.

Below we illustrate with numerical examples. Suppose the oil price can assume two different values, 1.5 and 0.5, with equal probability. The expected price is thus 1. Let the initial petroleum wealth be normalized as $w = 1$. The objective function becomes

$$\text{Max } EV = 1 - 0.5(1 - c)^2 \tag{7.13"a}$$
$$+ (1 + r)^{-1}0.5\{[1 - 0.5(1 - (1 - c)1.5)^2] + [1 - 0.5(1 - (1 - c)0.5)^2]\}$$

$$\text{Max } EV = 1nc + (1 + r)^{-1}0.5\{1n[(1 - c)1.5] + 1n[(1 - c)0.5]\} \quad (7.13''\text{b})$$

$$\text{Max } EV = -c^{1-z} + (1 + r)^{-1}0.5\{-[(1 - c)1.5]^{1-z} \quad (7.13''\text{c})$$
$$- [(1 - c)0.5]^{1-z}\}$$

and the first-order condition becomes

$$1 - c = (1 + r)^{-1}0.5\{[1 - (1 - c)1.5]1.5 + [1 - (1 - c)0.5]0.5\} \quad (7.14'\text{a})$$

$$c^{-1} = (1 + r)^{-1}0.5\{1.5[(1 - c)1.5]^{-1} + 0.5[(1 - c)0.5]^{-1}\} \quad (7.14'\text{b})$$

$$(1 - z)c^{-z} = (1 + r)^{-1}0.5\{1.5(1 - z)[(1 - c)1.5]^{-z} + 0.5(1 - z) \quad (7.14'\text{c})$$
$$[(1 - c)0.5]^{-z}\}.$$

To focus on the change induced by uncertainty, we may as well set $r = 0$. In that case, the wealth would be spread evenly over the two periods in the certainty case, so that $c = 0.5$ for all three functions. For the three functions Equation (7.14') implies the following:

Quadratic function	$c = 0.56$	$c_u > c_d$
Logarithmic function	$c = 0.5$	$c_u = c_d$
Constant elasticity function ($z = 2$)	$c = 0.46$	$c_u < c_d$

This numerical example confirms the results derived above. The graph of the three utility functions is shown in Figure 7.6; note that the logarithmic and the constant elasticity functions, which produce negative values for the interval $0 < c < 1$, have been displaced vertically so as to show identical utility for all three for $c = 1$, the initial petroleum wealth. The differences between these functions provides a better understanding as to why the extraction rate in period 1 is differently affected in these cases. The difference between the upside and downside risk is not very great for the quadratic function; it is not much more difficult to live with little wealth than it is pleasant to live with great wealth in period 2. In this case the extraction rate in period 1 is increased, to take advantage of the certainty with regard to wealth in that period. The constant elasticity function shows a much greater sensitivity toward downside risk; it is decidedly more unpleasant to live with little wealth than it is pleasant to live with great wealth in period 2. Hence, care is taken to ensure that not too little wealth is left in period 2, to make sure that there is sufficient consumption even in the case of low oil prices.

Figure 7.6
Three Popular Utility Functions

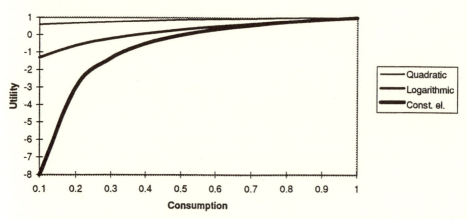

Note: $u = 1 - 0.5(1 - c)^2$ (Quadratic), $u = lnc$ (Logarithmic), and $u = -c^{1-z}$ (Constant Elasticity of Marginal Utility, z).

NOTES

1. This is the rate of discount applied by the Norwegian Ministry of Finance. It is probably well above the real rate of return on investment that can be achieved by investing both in the Norwegian economy and in financial markets abroad.

2. *Source*: IMF, quoted here from M. Burda and C. Wyplosz, *Macroeconomics* (Oxford: Oxford University Press, 1997), pp. 22–23.

3. Alaska Permanent Fund, 1996 Annual Report.

4. In real terms the fund income in 2010 will be a lot lower than the oil revenue of the state in 1981, as the U.S. dollar had lost at least 40 percent of its purchasing power when this projection of the fund's future income was made.

5. The maximum principle is explained in a number of textbooks on mathematical economics that cover dynamic optimization. See, e.g., textbooks by Chiang; Seierstad and Sydsæter; Kamien and Schwartz; or Leonard and Long.

6. On two-period models, see S. A. Lippman and J. J. McCall, "The economics of uncertainty," Chapter 6 in *Handbook of Mathematical Economics*, Vol. 1, eds., K. J. Arrow and M. D. Intriligator (Amsterdam: North-Holland, 1981).

7. There exists a large literature on this. Suffice it to say that risk aversion will be reflected in a concave utility function, so that $u'(c) > 0$ and $u''(c) < 0$.

Chapter 8

Petroleum Discoveries and Structural Changes

RESOURCE ENDOWMENTS: A MIXED BLESSING?

As alluded to earlier, natural resources are unevenly distributed around the globe. Some countries and areas are rich in natural resources while others are desolate. Changes in technology and resource discoveries sometimes make countries rich overnight. Saudi Arabia was once a poor province in the Ottoman empire and inhabited by nomads. Oil production began on a small scale in Saudi Arabia just before World War II and for a long time the country has been the foremost oil-exporting country of the world. The oil-exporting countries in the Middle East and elsewhere made a quantum leap in wealth during the two oil price hikes in the 1970s, but these were not sustained.

Nevertheless, countries that are rich in natural resources often are not manifestly better off than their neighbors. Norway is rich in petroleum deposits, hydroelectric power, and fish. Neighboring Denmark has much less oil and gas, no hydroelectric power, and shares overexploited fish stocks in the North Sea with her fellow European Union members. Yet the standard of living in Denmark is certainly comparable with Norway's.[1] Measured in a way that corrects for different purchasing power of national currencies the Danish gross domestic product (GDP) per capita was 11 percent higher than the Organization for Economic Cooperation and Development (OECD) average in 1995 while the Norwegian one was 17 percent higher. This is a smaller difference than the difference in natural resource endowments might lead one to expect.

One possible reason why resource-rich countries are not immensely richer than others is that resource-poor countries have to live by their wits while the resource rich can afford to relax. Whatever the reason, the low and perhaps inverse correlation between economic development and resource endowment is intriguing. We shall not attempt to resolve this question here but instead focus on the structural problems that typically accompany sudden discoveries of significant resource endowments, or what amounts to the same thing, a significant and sudden change in the value of known resources. For Norway both of these occurred at about the same time. Oil was discovered in the Norwegian sector of the North Sea just before Christmas in 1969, and the value of these finds was greatly enhanced, if only temporarily, by the two oil price hikes in the 1970s.

STRUCTURAL CHANGES AND "THE DUTCH DISEASE"

The problems of adjusting to sudden changes in natural resource endowments were much discussed in the late 1970s and early 1980s, in the wake of the oil price increases. In Europe it was particularly the Dutch economy that caught attention, and in fact the term "Dutch Disease" was invented for this class of problems.[2] The phrase caught on and even found its way into academic journals, although usually in quotes. The phrase is alleged to have been first put into print in a heading in *The Economist*, November 26, 1977 (pp. 82–83). The symptoms of the disease were stagnation in industrial production, falling corporate investment, and rising unemployment. Yet the Dutch currency was strong, and the balance of payments was positive. The root cause was identified as large exports of natural gas, pushing both the wage level and the exchange rate upward and reducing competitiveness, with government income from gas production fueling government spending.

But what is disease and what is just normal growth pains? That some structural changes should follow changes like resource discoveries is almost self-evident. To analyze these changes the economy is often divided into two sectors, a competitive sector, which sells "traded goods" (and services) abroad or at home in competition with foreign firms, and a sector sheltered from foreign competition. For goods and services sold on the market the critical distinction is between firms selling goods whose prices are determined in foreign markets and firms that can react to increases in domestic demand by raising their prices. Firms whose prices are determined in foreign markets cannot react to wage increases by raising their prices; they will have to accept cuts in their profits and, eventually, go out of business. Firms that do not face foreign competition are supposed to be able to react to wage increases by raising their prices, or perhaps doing so whether or not wages have increased. This notion is not without problems; formidable coordination problems must be overcome in order to raise prices in response to wage increases. If there are many firms competing domestically (such as in the construction industry), why should they not compete on price, just as in foreign markets? It may be more important, in the analysis to follow, that a substantial part of the sheltered sector consists of government services that are not subject to the discipline of the market.

The advantage of the two-sector model is that it lends itself readily to a

Figure 8.1
Change in the Production Possibilities as a Result of a Resource Discovery

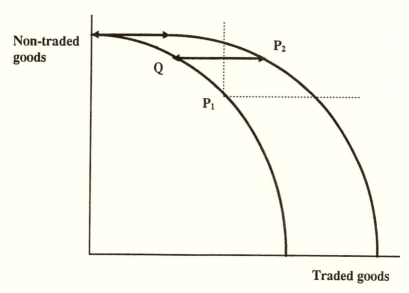

Note: The amount of traded goods that it is possible to produce and import increases, for any given amount of nontraded goods produced. Some of the increase in wealth is used to increase the consumption of nontraded goods, so production and consumption increase from P_1 to P_2. Hence the production of traded goods increases by less than the shift of the production possibility curve, implying that some "traditional" sectors must contract.

diagrammatic analysis. Figure 8.1 shows how much a country can produce of traded and nontraded goods with the productive resources at its disposal. Initially some of both will be produced and consumed, like at point P_1[3] Suddenly a large endowment of traded goods is discovered, or the price of some exported commodity increases substantially. The country will be able to produce a much greater value of exported goods than before, so the production possibility curve shifts to the right. But it is not very likely that the inhabitants of the country would desire to use all their new-found riches to buy traded goods. If people desire more of both traded and nontraded goods as they become richer, the new production will take place somewhere to the northeast of P_1, such as at P_2.

This, however, means that production in the "traditional" competitive sector must fall, for the production of traded goods does not increase by a value equal to the value of the new commodity (or the increase in the value of the commodity whose price has risen). And how is this to come about? Consider Figure 8.2. The width of the box shows the available labor in the economy. That labor is allocated to producing traded and nontraded goods. The amount used in the two sectors is measured from the two sides of the box, and the curves labeled *VMP* show the value of the marginal product of labor in producing traded and

Figure 8.2
Labor Market Adjustments to a Resource Discovery

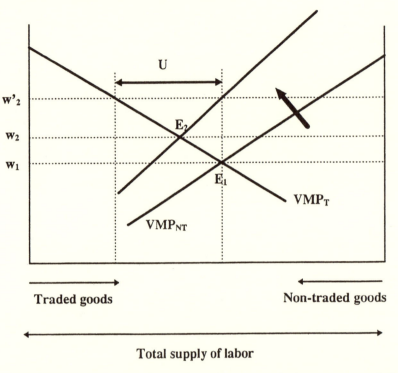

Note: The allocation of labor between two sectors of the economy is shown, a sector producing
traded goods and a sector producing nontraded goods, with the allocation of the available labor
(width of the box) to the traded goods sector being measured from the left, and vice versa for
the nontraded goods. VMP_{NT} and VMP_T show the value of the marginal product of labor in the
two sectors. In labor market equilibrium the wage rate (w) is equal to the VMP, and with a
common wage rate both VMPs are equal. Then a resource discovery leads to an upward shift
in the VMP for nontraded goods, because the price of these goods can be raised while the
prices of traded goods are set in foreign markets. Producers of nontraded goods will be able
to raise wages to w'_2 without reducing their production, but if production of nontraded goods
is to increase the wage rate must be lower than that. In the long term, with mobility of labor
between sectors, the wage rate will be w_2, with less production of "traditional," traded goods
and greater production of nontraded goods.

nontraded goods. We shall assume that the *VMP* curve for traded goods will
not be affected by the change, which amounts to saying that the new commodity
(or the one whose value has risen) can be produced without labor. This is, of
course, not true, but oil and gas and other mineral resources can usually be
produced with much less labor than industrial products, let alone services, so
the demand for labor arising from the new industry is of limited consequence.

An increase in the price of a traded commodity would not require an increase in the use of labor.

Initially the economy is in equilibrium at point E_1. With a uniform wage rate in both sectors (w_1) the *VMP* in both sectors will be equal. Then the increase in the value of traded goods comes along. Demand increases, for both traded and nontraded goods, but demand for traded goods increases by less than the production. To match demand and supply, the production of nontraded goods will have to increase, and the only way this can happen in a market economy is by raising the price of the nontraded goods relative to the traded goods, making it attractive to expand production of nontraded goods. The price of traded goods is given in foreign markets, so it is the price of nontraded goods that must rise. This moves the *VMP* curve for the nontraded goods upward, as the *VMP* is the product of the marginal physical product and the price of the commodity. The new equilibrium allocation of labor would be at point E_2, with more labor being allocated to nontraded goods and less to "traditional" traded goods. The wage rate will have increased, and some firms producing traded goods will have been put out of business.

A part of the change in relative prices will most likely be produced through changes in the exchange rate. An increase in exports means increased supply of foreign currency, which will lower its price or, what amounts to the same thing, raise the exchange rate of the domestic currency. An increase in the accumulation of foreign assets, such as investing a petroleum fund abroad, may counteract this however. A higher exchange rate means that the price of traded goods falls relative to the price of nontraded goods when expressed in domestic currency, which is precisely what happens in Figure 8.2.

So far the only consequences of the increased riches are normal and predictable; rather than being of any pathological nature they seem by and large desirable. Why should people want to use an increased income on food or gadgets and not on education, health care, entertainment, and improved housing?[4] Such changes in demand call forward changes in the composition of production, a process by which all will not win in the same proportion and some might lose. But could there be a disease, even if temporary and self healing?

Consider again Figure 8.2. A movement from E_1 to E_2 will take time and may be a rough ride. Labor probably cannot be moved quickly from the production of traded to the production of nontraded goods and services. There may have to be some retraining, and perhaps a new generation with new skills has to come to the fore. Maybe the first thing that happens is that the price of nontraded goods increases without a corresponding increase in production, through a rise in the exchange rate, or by domestic producers raising their prices in response to increased demand. What will union leaders do? Sit idly by and watch share prices and profits rise? Most likely they will demand their piece of the cake by demanding wage increases, which the producers of nontraded goods will be able to afford. But as wages rise in the nontraded sector the appetites

Figure 8.3
Value of Production of Crude Oil and Natural Gas in Norway (in Millions of "Kroner")

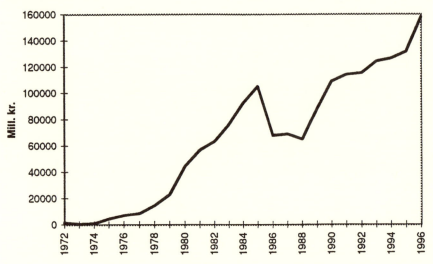

Source: Statistics Norway, *Oil and Gas Activity.*

of those who work in the competitive sector will be whetted. They will demand wages on par with the sheltered sector, and sometimes this will be facilitated by centralized unions demanding equal pay across sectors and firms. But the firms in the competitive sector cannot afford this, unless they cut their production. Wages might be pushed up to the level w'_2, resulting in an unemployment of U. Eventually the market will reach the new and lower equilibrium wage w_2, but it might take some time and in the meantime we would be left with the disease symptom of unemployment. This is particularly likely to happen if much of the revenue in the new industry ends up with the government, making it possible to be generous with social programs and unemployment benefits.

In the long run wages need not rise at all. If both industries are characterized by constant returns to scale and there is free movement of capital, the return on capital will be the same in all industries everywhere, and the wage rate will also be the same.[5] If people are able to migrate into the area where the booming sector is located, this will also tend to equalize wages. The natural resource industry is, however, not a constant returns to scale industry and rents in this industry will persist. Somehow the rents from this activity will be distributed and will make people in the resource-rich area richer than people elsewhere. If there is free movement of labor and every resident in the oil-producing area is entitled to a share in the oil rents, people will migrate to get a share in these rents, which will tend to equalize the income with the rest of the world. If the

Figure 8.4
Shares of Competitive and Sheltered Sectors in the GDP of Mainland Norway

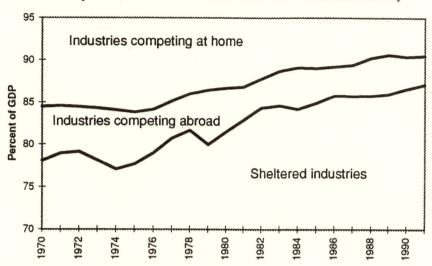

Source: Statistics Norway, *Historical Statistics 1994.*

governments of resource-rich areas try to distribute the rent through social programs and the like, they must be able to keep foreigners out if they are not willing to share the rents.

The structural changes discussed above are no less relevant in the other direction. When a resource-rich country runs out of its oil, or if its value drops precipitately, its income will shrink and demand will fall. This will hit both sectors of the economy. People must now be pushed out of the sheltered sector and into competitive industries. Perhaps it is here that the disease really sets in; it is not as easy to cut consumption as it is to increase it. If the production possibility curve in Figure 8.1 shifts back to its original path and people try to maintain the previous standard of living, consuming at P_2, they will run a deficit in their foreign trade equal to $P_2 - Q$. This is, of course, not sustainable. A reversibility problem of this kind can be seen in Figure 8.6 in the development of the balance of payments for Norway in the wake of the drop in oil prices in 1986.

Dutch Disease or not, it appears curable. The Netherlands, which rightly or wrongly gave name to this phenomenon, have lately been doing better than their neighbors. Their GDP increased more than that of Germany, France, and the United Kingdom between 1991 and 1996. Their consumer prices rose less than in Germany and the United Kingdom. Their unemployment rate is lower than in Germany, France, and the United Kingdom, and so is their public sector deficit as percent of GDP.[6]

Figure 8.5
Shares of Manufacturing and Government Services in the Labor Force in Norway and Sweden

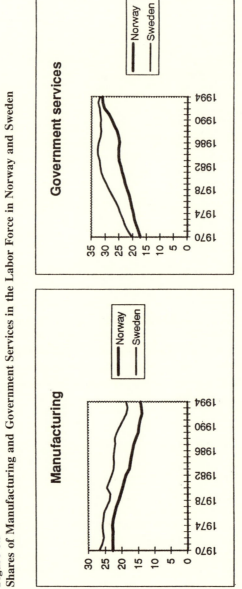

Sources: OECD, *National Accounts,* and, for Norway after 1991, Statistics Norway, *Yearbook of Statistics.*

Figure 8.6
Balance of Payments for Norway and Sweden, in Percent of Exports

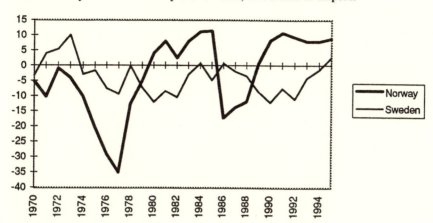

Source: OECD, *National Accounts.*

THE NORWEGIAN EXPERIENCE

Did Norway catch the Dutch Disease? The bacteria, or the virus, were certainly there. Figure 8.3 shows the development of Norway's oil and gas production since the petroleum era began. Oil production in Norway did not start on a significant scale until 1975. It then rose rapidly in value, dipped severely in 1986, but started to rise again in the late 1980s, mainly due to increased production. In 1996 Norway was the fifth largest oil-producing country in the world (after Saudi Arabia, the United States, Russia, and Iran).[7] Production of oil and gas amounted to 15 percent of GDP that year, exports of oil and gas were 38 percent of total exports, and oil and gas provided 16 percent of government revenue.

Some of the symptoms of the Dutch Disease seem to be in place in Norway. Figure 8.4 shows the breakdown of the GDP in "Mainland Norway" (i.e., excluding the petroleum industry and shipping) into three sectors, the sheltered sector and two competitive sectors, one that competes in foreign markets and one that meets foreign competition in its home market. The sheltered sector grew from just under 80 percent in the early 1970s to 87 in 1991.[8] Industries competing abroad shrank from 6 to 3 percent of GDP, and industries competing at home shrank from 15 to 10 percent of GDP.

But has there been a disease? If there had, we would expect the economic development in Norway to have departed radically from that of neighboring countries without oil and gas. To throw light on this question it seems reasonable to compare the development in Norway and Sweden. They share the Scandinavian peninsula, with a border of over 1,000 kilometers mostly coinciding with

Figure 8.7
**Net Savings of the Public Sector in Norway and Sweden, in Percent of Total
Receipts**

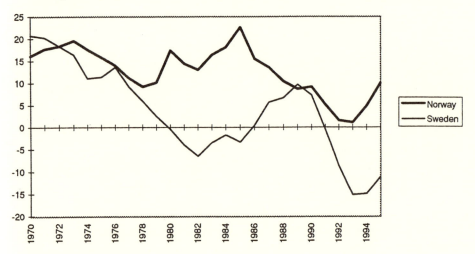

Source: OECD, *National Accounts.*

the mountain range that stretches along the peninsula from north to south. Their
culture and traditions are similar, and their languages are mutually intelligible.
One salient difference is size; there are about twice as many Swedes as there
are Norwegians. Another is resource endowments; despite considerable exploration effort the Swedes have not found any oil on their territory.

Figures 8.5 to 8.9 compare the economic development in Norway and Sweden
since 1970, well before the oil boom began in Norway. In both countries manufacturing, measured as its share of total employment, has declined at a fairly
similar rate (Figure 8.5). The employment in manufacturing is lower in Norway
than in Sweden, but this has always been the case. The share of government
services has risen in both countries. In Sweden this rise was fairly even up until
1984, but since then the share of the labor force employed in government services has stagnated and in fact declined periodically. In Norway the rise of employment in government services has been particularly rapid since 1988, but this
was well after the beginning of the oil boom and in fact after the reversal of
1986–87. Government services in Norway nevertheless still claim a slightly
lower share of the labor force than the case is in Sweden.

The balance of payments in Norway has had a rickety ride, largely as a result
of the activity in the petroleum industry (Figure 8.6). Investments in oil platforms, pipelines, and so forth were heavy in the 1970s, while the revenues did
not start to flow with great force until the 1980s. The balance of payments was
therefore strongly negative, reaching a low point in 1977. The years of high oil

Figure 8.8
Unemployment in Norway and Sweden, Measured in Percent of the Labor Force

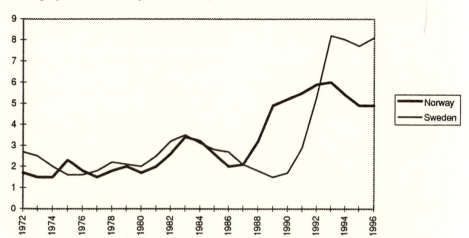

Sources: Statistics Norway, *Yearbook of Statistics*, and the Central Bureau of Statistics, Sweden, *Yearbook of Statistics.*

prices in the early 1980s produced a surplus on the current account, but the steep fall of oil prices generated a corresponding deficit in 1986–88. Increased production and a partial recovery of oil prices turned the current account around again into surplus in 1990. The Swedish current account has almost consistently been in deficit since 1974, with a corresponding accumulation of foreign debt.

There are also differences in government finances. Net savings in the Norwegian public sector have consistently been positive, but took a severe dip in the recession years of 1992–93 (Figure 8.7). In Sweden the savings were negative, but not by much, in 1981–85, but turned negative by major amounts in the early 1990s, leading to an unsustainable accumulation of government debt.

Both Sweden and Norway were the envy of many others for years because of the low unemployment rate (Figure 8.8). Unemployment hovered between 2 and 3 percent of the labor force and showed a rather similar pattern in both countries. Unemployment in Norway shot up in the late 1980s and reached 6 percent in 1993 but is slowly coming down. Unemployment in Sweden rose in the early 1990s and reached 8 percent and has stayed high ever since. The employment history of both countries thus is broadly similar, with the exception of the most recent years when Swedish unemployment seems to have become stuck at a historically high rate. Neither is there a significant difference in labor force participation in the two countries; the labor force participation rate of people fifteen to sixty-four years old grew from 62.5 percent in 1974 to 64.2 percent in 1994 in Norway, and fell slightly in Sweden, from 64.5 to 63.7.[9] Norwegians do not seem to be floating on oil in early retirement while Swedes have to toil to earn their living.

Figure 8.9
GDP per Capita, Measured at Purchasing Power Parity, in Percent of the OECD Average

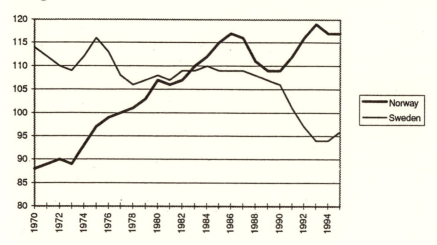

Source: OECD, *National Accounts.*

What are we to make of this? The economic development in Norway and Sweden seems remarkably similar, in terms of both structural changes and changes in unemployment. If the Norwegian economy is afflicted by disease it does not appear to be the Dutch one but perhaps a common Scandinavian malaise, as the symptoms in Sweden seem quite similar. The oil does make a tremendous difference, however, for external debt and government debt. While Sweden is fighting an uphill struggle against both, the high export and tax revenues from oil and gas mean that neither external nor internal debt are constraints on the Norwegian economy. The difference also shows in the development of the GDP (Figure 8.9). Sweden used to be the envy of its Nordic neighbors and many others for its high standard of living. In 1970 the Swedish GDP per capita was well above the OECD average while the Norwegian one was equally far below. Since then Sweden has declined in relative terms while Norway has risen; by the early 1980s the GDP per capita was about the same in both countries. In the 1990s the gap has widened quickly; while Sweden has stagnated the Norwegian economy has grown. Sweden has now fallen below the OECD average in terms of GDP per capita while Norway is well above that average.

NOTES

1. It is somewhat ironic in this context that Denmark got a disproportionately small and unproductive share of the North Sea when it was carved up in the 1960s. That division was probably made easier because no one knew what was underneath the seabed,

and expectations were generally gloomy. The first oil field that was discovered in the Norwegian sector, Ekofisk, lies straight west of Jutland but Norway got the area because of the equidistance principle, which draws boundaries halfway between the nearest lands. Still, the equidistance principle was never followed automatically; the parties involved had to negotiate and agree on their sea boundaries. A story that refuses to die has it that the Danish foreign minister was under the influence of alcohol during the final negotiations and was eager to conclude them. One version of the story holds that the booze was bought by the Norwegian foreign ministry. If true, the bottle would rank among the better investments in history.

2. One of the classics on these problems is Corden (1984), which has a long list of references.

3. Note that, in technical terms, production and consumption take place at point P_1. This is so because both exports and imports are treated as one and the same commodity and exports are assumed to be equal in value to imports.

4. Education, health care, and entertainment are not totally nontradable services, but the bulk of these typically are. Housing is not totally nontradable either; prefabricated houses can be and are imported in many countries.

5. For a production function that is homogeneous of degree one the marginal productivity depends only on the ratio of capital to labor (in the two-factor case). If the rate of return on capital is the same in all countries and all use the same technology, the capital-labor ratio will be the same everywhere for traded goods. Therefore, the marginal productivity of labor must be the same everywhere, and since the prices are the same everywhere, so must the wage rates.

6. Quoted from the *Financial Times*, February 27, 1997, p. 10.

7. *BP Statistical Review of World Energy*, 1996.

8. Unfortunately these figures, provided by Statistics Norway: *Historical Statistics 1994*, have not been updated for later years.

9. OECD, *Historical Statistics 1960–1994* (Paris, 1996).

Bibliography

Adelman, M. A. (1972). *The World Petroleum Market*. Baltimore: Johns Hopkins University Press.

———. (1990). Mineral depletion, with special reference to petroleum. *Review of Economics and Statistics* 72, pp. 1–10.

———. (1995). *The Gene out of the Bottle*. Cambridge, Mass.: MIT Press.

Austvik, O. G. (1997). Gas pricing in a liberalized European market; will the rent be taxed away? *Energy Policy* 25, pp. 997–1012.

Bjerkholt, O., E. Offerdal and, S. Strøm, eds. (1985). *Olje og gass i norsk økonomi* (oil and gas in the Norwegian economy). Oslo: Unversitetsforlaget.

Bjerkholt, O., E. Gjelsvik, and Ø. Olsen (1990a). The Western European gas market: deregulation and supply, in Bjerkholt, O., Ø. Olsen, and J. Vislie (eds.), pp. 3–28.

Bjerkholt, O., Ø. Olsen, and J. Vislie, eds. (1990b). *Recent Modelling Approaches in Applied Energy Economics*. London: Chapman & Hall.

BP Statistical Review of World Energy (1998 and earlier years). The British Petroleum Company.

Bradley, R. L., Jr. (1995). *Oil, Gas, and Government: The U.S. Experience*. Lanham, Md.: Rowman & Littlefield.

Chiang, A. C. (1992). *Elements of Dynamic Optimization*. New York: McGraw-Hill.

Corden, W. M. (1984). Booming sector and Dutch Disease economics: Survey and consolidation. *Oxford Economic Papers* 36, pp. 359–380.

DeVany, A. S., and W. D. Walls (1995). *The Emerging New Order in Natural Gas*. Westport, Conn.: Quorum Books.

Estrada, J., A. Moe, and K. D. Martinsen (1995). *The Development of European Gas Markets*. Chichester: John Wiley & Sons.

Garnaut, R., and A. C. Ross (1975). Uncertainty, risk aversion and the taxing of natural resource projects. *Economic Journal* 85, pp. 272–287.

Griffin, J. M., and D. J. Teece, eds. (1982). *OPEC Behavior and World Prices*. London: Allen & Unwin.

Groot, F., C. Withagen, and A. de Zeeuw (1992). Note on the open-loop von Stackelberg equilibrium in the cartel versus fringe model. *Economic Journal* 102, pp. 1478–1484.

Hoel, M., B. Holtsmark, and J. Vislie (1990). The European gas market as a bargaining game, *in* Bjerkholt, O., Ø. Olsen, and J. Vislie (eds.), pp. 49–66.

Jensen, J. T. (1992). Open access—the new market approach to natural gas policy. *Energy Policy* (October): 1005–1014.

Jones, P. E. (1988). *Oil. A Practical Guide to the Economics of World Petroleum*. Cambridge, Mass.: Woodhead-Faulkner.

Kamien, M. I., and N. L. Schwartz (1991). *Dynamic Optimization*. Amsterdam: North-Holland.

Kenney, J. F. (1996). Impending shortages of petroleum re-evaluated. *Energy World* (June): 16–18.

Kenney, J. F. (1997). ''Reply to 'abiotic origins of oil-revisited (sic).' '' Unpublished manuscript, Joint Institute of the Physics of the Earth, Russian Academy of Sciences, Moscow; Gas Resources Corporation, Houston.

Kretzer, Ursula M. H. (1993). Allocating oil leases. *Resources Policy* (December): 299–311.

Leonard, D., and N. V. Long (1992). *Optimal Control Theory and Static Optimization in Economics*. Cambridge: Cambridge University Press.

Libecap, G. (1989). *Contracting for Property Rights*. Cambridge: Cambridge University Press.

Maddison, A. (1991). *Dynamic Forces in Capital Development*. Oxford: Oxford University Press.

McCray, A. W. (1975). *Petroleum Evaluations and Economic Decisions*. Englewood Cliffs, N.J.: Prentice-Hall.

Moran, T. (1982). Modeling OPEC behavior: Economic and political alternatives, *in* Griffin, J. M., and D. J. Teece (eds.).

Nash, J. F. (1950). The bargaining problem. *Econometrica* 18, pp. 155–162.

Newbery, D. M. (1981). Oil prices, cartels, and the problem of dynamic inconsistency. *Economic Journal* 91, pp. 617–646.

———. (1991). The open-loop von Stackelberg equilibrium in the cartel versus fringe model: A reply. *Economic Journal* 102, pp. 1485–1487.

Nystad, A. N. (1985). Petroleum taxes and optimal resource recovery. *Energy Policy* (August): 381–401.

———. (1987). Rate sensitivity and the optimal choice of production capacity of petroleum reservoirs. *Energy Economics* (January): 37–45.

———. (1988a). Petroleum reservoir management: A reservoir economic approach. *Natural Resource Modeling* 2, pp. 345–382.

———. (1988b): On the economics of improved oil recovery: The optimal recovery factor from oil and gas reservoirs. *The Energy Journal* 9, pp. 49–61.

OECD (1997). *Historical Statistics 1960–95*. Paris: Organisation for Economic Cooperation and Development/International Energy Agency.

OECD/IEA (1994). *Natural Gas Transportation*. Paris: Organisation for Economic Cooperation and Development/International Energy Agency.

Seierstad, A., and K. Sydsæter (1987). *Optimal Control Theory with Economic Applications*. Amsterdam: North-Holland.

Shell International Exploration and Production (1997). Abiotic origins of oil-revisited. *Energy World* (March): 14.

Stern, J. P. (1990). *European Gas Markets*. Dartmouth: The Royal Institute of International Affairs.

Stoppard, M. (1996). *A New Order for Gas in Europe?* Oxford: Oxford Institute for Energy Studies.

Van Meurs, A. P. H. (1981). *Modern Petroleum Economics*. Ottawa: Van Meurs and Associates.

Vislie, J. (1990). Bargaining, vertical control, and (de)regulation in the European gas market, in Bjerkholt, O., Ø. Olsen, and J. Vislie (eds.), pp. 67–86.

Index